Test Bank to Accompany

Finite Mathematics

An Applied Approach

Eighth Edition

Abe Mizrahi
Indiana University Northwest

Michael Sullivan
Chicago State University

Prepared by

Henry Smith
Southeastern Louisiana State University

John Wiley & Sons, Inc.
New York / Chichester / Weinheim / Brisbane / Singapore / Toronto

To order books or for customer service call 1-800-CALL-WILEY (225-5945).

ISBN 0-471-35510-0

Printed in the United States of America

10 9 8 7 6 5 4 3

Printed and bound by Odyssey Press, Inc.

CONTENTS

FOREWORD

Test Bank to accompany ***Finite Mathematics: An Applied Approach***, 8th edition by Abe Mizrahi and Michael Sullivan contains sample test items reflecting the material found in the text. The test items are free-form questions and are listed by chapter.

Following the questions, the worked-out solutions to each item are given along with the particular section of the chapter the item comes from. The solution should provide an idea of the complexity and time requirements of the item.

Many of the problems can be worked using calculators and graphing calculators. Those that are especially calculator-active have been designated by an icon (🖳). Some problems not designated with an icon can become trivialized when a graphics calculator is used to solve them. No specific brand or model of graphing calculator is suggested. The solutions provided in this test bank were found using a Texas Instruments TI-82 and a Hewlett-Packard 48GX. All decimal approximations have been carried out to at least three places where appropriate.

The entire Test Bank has been incorporated into MICROTEST®, a microcomputer test preparation system, by Delta Software. This software has been designed to retrieve the questions in the Test Bank and print your tests and answer keys using a PC running Windows 3.1, 95, 98, or higher. Also available is an Apple Macintosh version of the software, running on System Tools 5.0 or higher. The questions on your tests printed by the software will be exact duplicates of those shown here, including the tables and illustrations. The software includes a host of powerful features and flexibility for selecting questions and producing multiple versions of tests without your having to retype the items. You can also add your own questions to the Test Bank. To obtain the MICROTEST® software for this textbook, contact your John Wiley sales representative.

CHAPTER 1

LINEAR EQUATIONS

1. Plot each point in the xy-plane. Tell in which quadrant each point is located:
   a. $(-2, 3)$ d. $(0, 6)$
   b. $(-1, -1)$ e. $(-6, 0)$
   c. $(1, 4)$ f. $(2, -4)$

2. Complete the following table if $2x + y = 8$. Use your points to graph the equation.

x	0		2		−3	4
y		0		−1		

For problems 3 – 8, find the slope of the line that is defined in each case.

3.

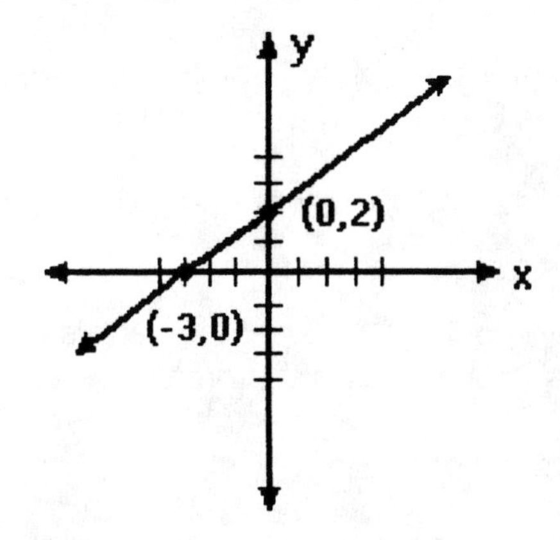

4. The line that passes through points $(2, 1)$ and $(4, -6)$.

5. $2x + y = 4$

6. $y = -3x + 1$

7. $x = 4$

8. $y = -1$

In problems 9 – 13, find a general equation for the line having the given properties.

9. Slope = 4; contains $(-3, 4)$

10. Contains the two points $(-2, 3)$ and $(1, 2)$

11. Slope = −2; y-intercept = $(0, 4)$

12. Slope undefined; contains the point $(1, 4)$

13. Parallel to the line $2x - y = 5$ and passing through $(-1, 2)$

14. Find the point of intersection of the lines $x + y = 5$ and $3x - y = 7$. Graph the pair of lines.

15. Find the slope and y-intercept of the line $2x - 3y = 6$

16. If they intersect, find the point of intersection of the lines $2x + 3y + 6 = 0$ and $-2x + 5y + 10 = 0$. If they do not intersect, state why.

17. Without solving, determine whether the lines below are perpendicular, coincidental, or parallel.

    L: $-2x + 3y + 8 = 0$
    M: $2x - 3y + 6 = 0$

18. Find the equation for the line that is perpendicular to the line $6x - 2y = 5$ and passing through (2, 1).

19. Find the point of intersection of the lines $2x - y = 4$ and $4x - 3y = -1$.

20. Find the point of intersection of the lines $2x + 3y + 6 = 0$ and $-2x + 5y + 10 = 0$.

21. A bank loaned $15,000, some at an annual rate of 16% and some at an annual rate of 10%. If the income from these loans was $1800, how much was loaned at 10%?

22. If $400 is borrowed at a simple interest rate of 14% per annum, what amount is due after 18 months?

23. A manufacturer produces items at a daily cost of $1.25 per item and sells them for $2 per item. The daily operational overhead is $450. What is the break-even point?

24. Graph the line $3x - 2y = 6$. Label the intercepts.

25. Graph the pair of lines $x - y = 4$, $2x + y = 6$. Label the y-intercepts and the point of intersection, if any.

26. Find the equation of the line shown:

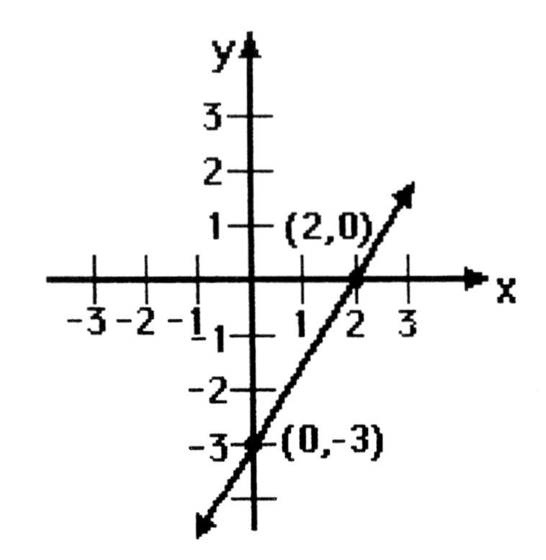

27. Give the equation of the line(s) shown below whose slope is negative.

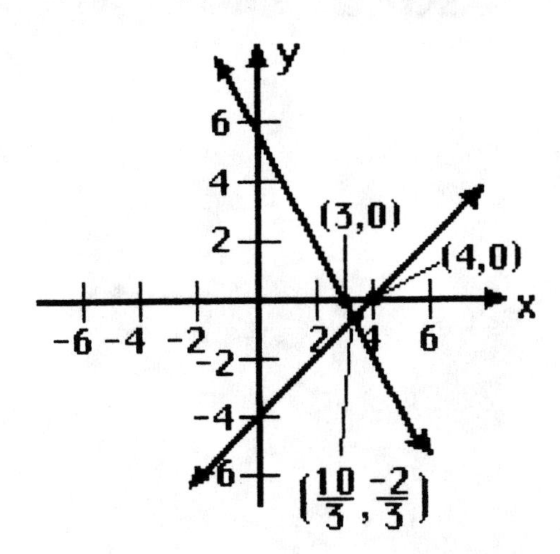

28. Given lines L_1 and L_2, with indicated slopes, identify the lines as either parallel, perpendicular, or intersecting but not perpendicular.

a. L_1: 2/3
   L_2: –3/2

b. L_1: 1/2
   L_2: 2

c. L_1: 2/5
   L_2: 2/5

29. Suppose the sales of a company are given by

$S(x) = \$200x + \$20{,}000$ where x is measured in years and x = 0 corresponds to the year 1996.

a. Find S when x = 3

b. Find the predicted sales in the year 2002 assuming the trend continues.

30. One solution is 10% acid and another is 22% acid. How many cubic centimeters of each should be mixed to obtain 100 cubic centimeters of a solution that is 12% acid?

31. For a certain commodity, the supply and demand equations have been estimated by S = 4p + 10 and D = –0.5p + 19. Find the market price.

32. For a certain commodity, the demand equation is given by D = –5p + 50. At a price of $1, twelve units of the commodity are supplied. If the supply equation is linear and the market price is $4, find the supply equation.

SOLUTIONS

1. (Section 1.1)

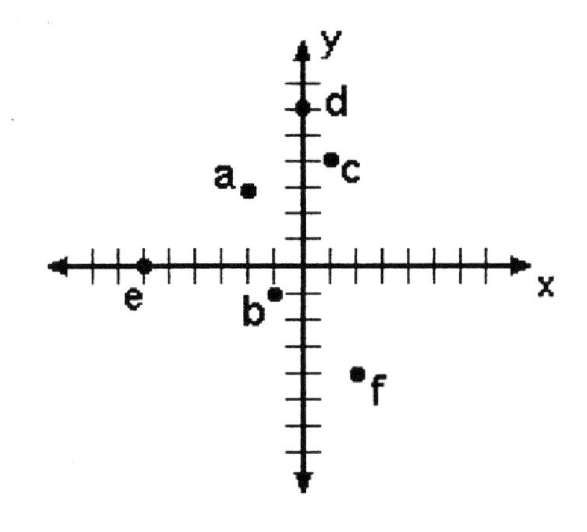

a. (−2, 3) QII
b. (−1, −1) QIII
c. (1, 4) QI
d. (0, 6) on y-axis
e. (−6, 0) on x-axis
f. (2, −4) QIV

2. (Section 1.1)

x	0	4	2	4.5	−3	4
y	8	0	4	−1	14	0

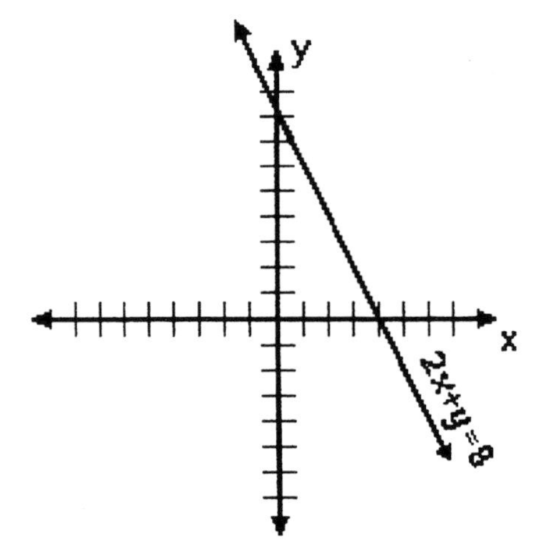

3. (Section 1.1)
$$\frac{2-0}{0-(-3)} = \frac{2}{3}$$

4. (Section 1.1)
$$\frac{-6-1}{4-2} = \frac{-7}{2}$$

5. (Section 1.1)
$y = -2x + 4$ so the slope is -2

6. (Section 1.1)
slope $= -3$

7. (Section 1.1)
a vertical line has no slope or an undefined slope

8. (Section 1.1)
a horizontal line has 0 slope

9. (Section 1.1)
$y - 4 = 4(x + 3)$
$4x - y = -16$

10. (Section 1.1)

$$\text{slope} = \frac{3-2}{-2-1} = \frac{-1}{3}$$

$$y - 2 = \frac{-1}{3}(x-1)$$

$3y - 6 = -x + 1$

$x + 3y = 7$

11. (Section 1.1)

$y = -2x + 4$

$2x + y = 4$

12. (Section 1.1)
$x = 1$

13. (Section 1.2)
slope = 2
$y - 2 = 2(x + 1)$
$2x - y = -4$

14. (Section 1.2)
$y = -x + 5$
$y = 3x - 7$
$-x + 5 = 3x - 7$
$12 = 4x$; $3 = x$
$y = -3 + 5 = 2$ (3, 2)

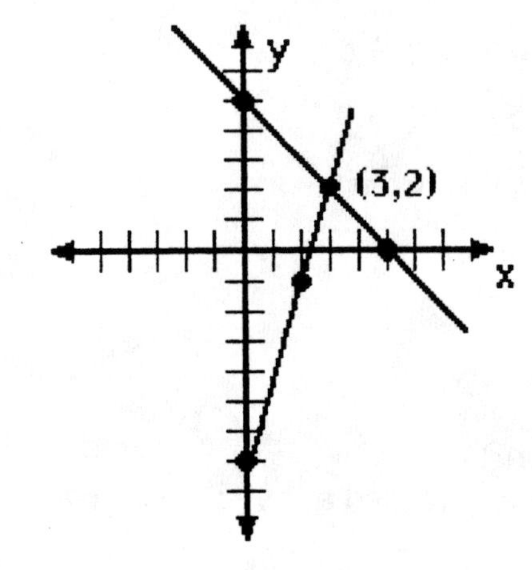

15. (Section 1.1)
$-3y = -2x + 6$
$y = 2x/3 - 2$
slope = 2/3; y-intercept = (0,–2)

16. (Section 1.2)

$$\begin{cases} 2x + 3y + 6 = 0 \\ -2x + 5y + 10 = 0 \end{cases}$$

$8y + 16 = 0$

$y = -2$, $2x - 6 + 6 = 0$

$x = 0$, $y = -2$ (0, –2)

17. (Section 1.2)

L: $3y = 2x - 8 \qquad y = \frac{2}{3}x - \frac{8}{3}$

M: $3y = 2x + 6 \qquad y = \frac{2}{3}x + 2$

equal slopes; unequal y-intercepts
The lines are parallel.

18. (Section 1.2)
If perpendicular to $6x - 2y = 5$, then the line must have a slope of $-1/3$ and it contains point (2, 1). Therefore, $y - 1 = -1/3\ (x - 2) \Rightarrow x + 3y = 5$

19. (Section 1.2)

$2x - y = 4$

$4x - 3y = -8$

$$\begin{aligned} y &= 2x - 4 \\ 4x - 3(2x - 4) &= -8 \\ -2x &= -20 \\ x &= 10 \\ y &= 20 - 4 = 16 \end{aligned}$$

$x = 10$; $y = 16$

20. (Section 1.2)

$$\begin{aligned} 2x + 3y + 6 &= 0 \\ -2x + 5y + 10 &= 0 \\ 8y + 16 &= 0 \end{aligned}$$

$y = -2$

$2x - 6 + 6 = 0$

$x = 0$, $y = -2$

21. (Section 1.3)
x = amount loaned at 16%; y = amount loaned at 10%

$$\begin{aligned} x + y &= 15000 \\ .16x + .1y &= 1800 \end{aligned}$$

$$\begin{aligned} 1.6x + y &= 18000 \\ -0.6x &= -3000 \end{aligned}$$

$x = 5000$, $y = 10000$

$10,000 was loaned at 10%

22. (Section 1.3)
$I = Prt = 400(.14)(1.5) = 84$ Amount due = 400 + 84 = $484

23. (Section 1.3)

x = number of items produced; R = revenue; C = cost

$$R = 2x;\quad C = 1.25x + 450$$
$$2x = 1.25x + 450$$
$$.75x = 450$$
$$x = 600$$

The break-even point is (600,1200).

24. (Section 1.1)

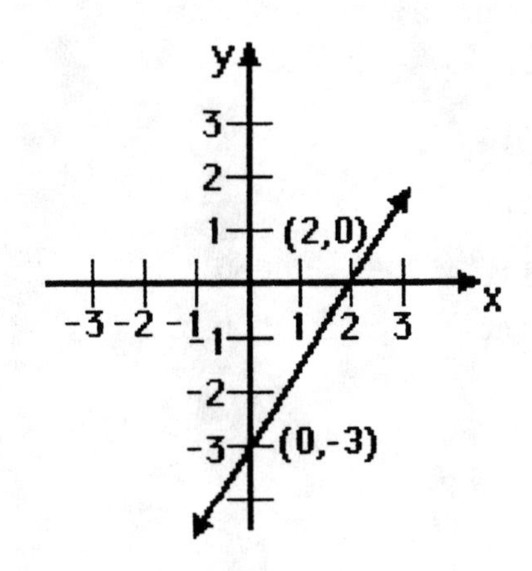

25. (Section 1.2)

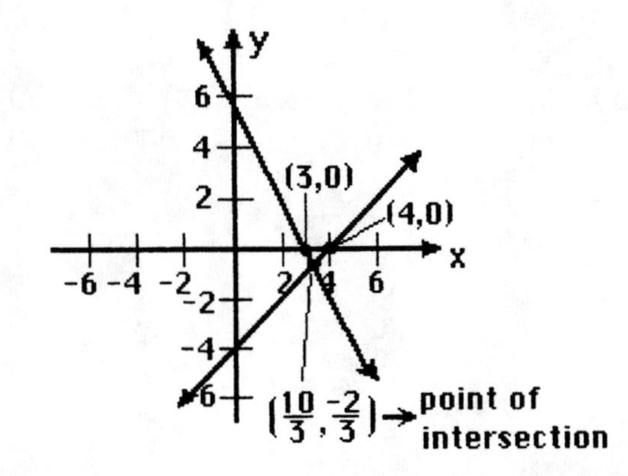

26. (Section 1.1)

slope = 3/2

Equation of the line is $y = \frac{3}{2}x - 3$

27. (Section 1.2)

The line through (3,0) and (0,6) has a slope of –2.
Equation of the line is $y = -2x + 6$

28. (Section 1.2)

a. perpendicular since $\frac{2}{3} \bullet \frac{-3}{2} = -1$

b. intersecting but not perpendicular since $\frac{1}{2} \bullet 2 \neq -1$ and $\frac{1}{2} \neq 2$

c. parallel since slopes are identical

29. (Section 1.3)

a. $S(3) = \$20{,}600$

b. If the year needed is 2002, then $x = 7$.

$S(6) = \$21{,}200$

30. (Section 1.3)

Let x = amount of 10% acid solution.

Then $100 - x$ = amount of 22% solution.

$.10x + .22(100 - x) = .12(100)$

$x = 83\ 1/3$ cc

Therefore, you need 16 2/3 cc of 22% solution.
83 1/3 cc of 10% solution.

31. (Section 1.3)

$4p + 10 = -0.5p + 19$

$4.5p = 9$

$p = 2$

32. (Section 1.3)

The points are (12, 1) and (x, 4) where $x = -5(4) + 50 = 30$

$$m = \frac{4 - 1}{30 - 12} = \frac{1}{6}$$

$$p - 1 = \frac{1}{6}(S - 12)$$

$6p - 6 = S - 12$

$S = 6p + 6$

CHAPTER 2

SYSTEMS OF LINEAR EQUATIONS; MATRICES

1. Is (1,–1) a solution to:

$$\begin{cases} 5x - y = 6 \\ 2x + y = 2 \end{cases} \quad ?$$

2. Solve by substitution or elimination.

$$\begin{cases} 3x + y = 4 \\ x - 3y = -2 \end{cases}$$

In problems 3 – 5, write the augmented matrix of each system.

3. $3x + y = 4 \qquad x - 3y = 8$

4. $x - y + 2z = 4 \qquad 3y + z = 5 \qquad 3x - y = 4$

5. $2x_1 - 3x_2 + x_3 = 0 \qquad x_1 - x_2 - 2x_3 = -2 \qquad -5x_1 + x_2 = -4$

6. Perform the row operation $R_2 = (-2)r_1 + r_2$ on the matrix

$$\begin{bmatrix} 1 & 5 & -6 \\ 6 & -1 & 2 \end{bmatrix}$$

In problems 7 – 20, determine whether each system has a unique solution, no solution, or infinitely many solutions. If a solution exists, write it down.

7. $x + 2y = 5 \qquad x - y = -1$

8. $x + y + 2z = 6 \qquad -x - y + z = -3 \qquad 2x + y + 3z = 7$

9. $x_1 + 2x_2 + 10x_4 = 17 \qquad x_1 + 3x_2 + 14x_4 = 25$
   $2x_1 + 2x_2 - x_3 + 10x_4 = 14 \qquad 2x_1 + x_2 + 10x_4 = 12$

10. $x + y + z = 3 \qquad x + 2y + z = 0 \qquad x + 3y + z = 1$

11. $x - y = -2 \qquad 3x - 2y = -2 \qquad 5x - 3y = -2$

12. $x_1 + x_2 + x_3 = 3 \qquad 2x_1 - x_2 - 2x_3 = -3 \qquad 3x_2 - 4x_3 = -3$

13. $y = x \qquad x + y = 4$

14. $x + y = 1 \qquad x - y = 7$

15. $3x = 2y - 4 \qquad 6x - 4y = -4$

16. $x = \dfrac{2y}{3} \qquad y = 4x + 5$

17. $x_1 + 2x_2 - x_3 = 2$ $2x_1 + 2x_2 - 3x_3 = 1$

18. $x_3 + x_4 = 7$ $-x_1 + x_2 + x_4 = 5$
$x_1 + x_2 + x_3 - x_4 = 6$ $-x_2 + x_4 = 10$

19. $x_1 + x_2 = 5$ $x_2 - x_3 = 8$ $x_1 + x_3 = -2$

20 $x_1 + 2x_3 - x_4 = 5$ $2x_1 + x_2 + 3x_3 = 7$ $5x_1 + x_2 + 9x_3 - 3x_4 = 12$

21. The difference of two numbers is twelve. Three times the smaller one is four less than twice the larger. Find the smaller number.

In problems 22 – 24, the reduced row-echelon form of the augmented has one solution, no solution, or infinitely many solutions. State which.

22. $$\left[\begin{array}{ccc|c} 1 & 0 & 0 & -1 \\ 0 & 1 & 0 & 0 \\ 0 & 0 & 1 & 3 \end{array}\right]$$

23. $$\left[\begin{array}{ccc|c} 1 & 0 & 0 & 2 \\ 0 & 1 & 0 & 3 \\ 0 & 0 & 0 & 0 \end{array}\right]$$

24. $$\left[\begin{array}{ccc|c} 1 & 0 & 0 & 0 \\ 0 & 1 & 0 & 0 \\ 0 & 0 & 0 & 1 \end{array}\right]$$

25. What is the dimension of the matrix shown below?

$$\begin{bmatrix} 2 & 1 & 4 \\ 1 & 3 & 5 \\ -2 & 0 & 1 \\ 6 & 0 & 1 \end{bmatrix}$$

26. What is the dimension of the matrix shown below?

$$\begin{bmatrix} 1 & 3 & 5 \\ -2 & 0 & 1 \end{bmatrix}$$

27. Find x and y so that

$$\begin{bmatrix} x & 2y \\ 4 & 0 \end{bmatrix} = \begin{bmatrix} 3x+5 & y-2 \\ 4 & 0 \end{bmatrix}$$

28. Find x and y so that:

$$\begin{bmatrix} x+y & 4 \\ 8 & 0 \end{bmatrix} = \begin{bmatrix} 8 & 4 \\ x-y & 0 \end{bmatrix}$$

29. Find:

$$\begin{bmatrix} 2 & -6 & 0 \\ 4 & 2 & 3 \end{bmatrix} + \begin{bmatrix} 4 & 0 & -3 \\ 1 & 2 & 8 \end{bmatrix}$$

30. Find:

$$4\begin{bmatrix} 1 & -3 & 1 \\ 0 & 2 & 1 \end{bmatrix} - \begin{bmatrix} 8 & -1 & 1 \\ 2 & 0 & 1 \end{bmatrix}$$

31. If A is a matrix of dimension 2 x 3 and B is a matrix of dimension 3 x 5, is AB defined? If so, what is its dimension? Is BA defined? If so, what is its dimension?

32. Find:

$$\begin{bmatrix} 1 & 3 \\ 2 & -1 \end{bmatrix} \begin{bmatrix} 4 & 3 \\ 0 & 1 \end{bmatrix}$$

33. Find:

$$\begin{bmatrix} 4 & 3 \\ 0 & 1 \end{bmatrix} \begin{bmatrix} 1 & 0 & 2 \\ 2 & -1 & 0 \end{bmatrix}$$

34. Find:

$$\begin{bmatrix} 2 & -1 \\ 3 & -2 \end{bmatrix} \begin{bmatrix} 2 & -1 \\ 3 & -2 \end{bmatrix}$$

35. Find the inverse of:

$$\begin{bmatrix} 2 & 3 \\ -1 & -2 \end{bmatrix}$$

36. Find the inverse of:

$$\begin{bmatrix} 1 & 2 & 1 \\ 1 & 1 & 2 \\ 2 & 0 & 2 \end{bmatrix}$$

37. Show that the matrix

$$\begin{bmatrix} 8 & -4 \\ 4 & -2 \end{bmatrix}$$ has no inverse.

38. Find the inverse of:

$$\begin{bmatrix} 2 & 1 & 1 \\ 0 & 1 & 2 \\ 2 & 2 & 1 \end{bmatrix}$$

Use it to solve the following system of equations:

$$\begin{cases} 2x_1 + x_2 + x_3 = 4 \\ x_2 + 2x_3 = 0 \\ 2x_1 + 2x_2 + x_3 = 1 \end{cases}$$

39. A chemistry laboratory has available three kinds of hydrochloric acid (HCl): 5%, 15%, and 30% solutions. How many liters of each kind should be mixed to obtain 100 liters of 25% HCl? Provide a table showing at least three of the possible solutions.

40. Find the relative wages of each person for the closed input-output matrix:

$$\begin{array}{c} \\ A \\ B \\ C \end{array} \begin{array}{c} \begin{array}{ccc} A & B & C \end{array} \\ \begin{bmatrix} 1/3 & 1/2 & 1/3 \\ 1/3 & 1/4 & 1/3 \\ 1/3 & 1/4 & 1/3 \end{bmatrix} \end{array}$$

Take the wages of C as parameters and use x_3 = C's wages = $20,000.

41. Using the correspondence:

A → 1	B → 2	C → 3	D → 4	E → 5	F → 6	G → 7
H → 8	I → 9	J → 10	K → 11	L → 12	M → 13	N → 14
O → 15	P → 16	Q → 17	R → 18	S → 19	T → 20	U → 21
V → 22	W → 23	X → 24	Y → 25	Z → 26		

Encode the message "TEST IT" in triplets of numbers using the matrix:

$$\begin{bmatrix} 1 & 13 & 2 \\ 0 & -1 & 0 \\ 0 & 5 & 1 \end{bmatrix}$$

42. Compute A^T if $A = \begin{bmatrix} 2 & 3 & 0 \\ 1 & -1 & 5 \end{bmatrix}$

43. Find the least squares line of best fit to the data:

x	2	4	6	8
y	8	10	11	15

In problems 44 – 51, give the dimensions of the resulting matrix. Write "not possible" if the problem cannot be worked. Do *not* perform the operation.

$$A = \begin{bmatrix} 2 & 1 \\ -1 & 4 \end{bmatrix} \quad B = \begin{bmatrix} 2 & -1 \\ 4 & 0 \end{bmatrix} \quad C = \begin{bmatrix} 5 & 2 & 3 \\ 1 & 6 & -1 \end{bmatrix} \quad D = \begin{bmatrix} 3 & 1 \\ 2 & -1 \\ 4 & 1 \end{bmatrix}$$

44. AC

45. BD

46. inverse of B (B^{-1})

47. inverse of C (C^{-1})

48. transpose of A (A^T)

49. transpose of D (D^T)

50. (2A – 3B)C

51. CD

52. Use a graphing calculator and find the inverse of

$$\begin{bmatrix} 14 & -31 & 81 \\ 61 & -21 & 0 \\ 14 & -15 & -52 \end{bmatrix}$$

Approximate each entry to the nearest thousandth.

53. Use a graphing calculator and solve

$$\begin{cases} 21x + 3y - 20z = 5 \\ 18x - 4y + 31z = -7 \\ 10x - 51y + 13z = 110 \end{cases}$$

Approximate each entry to the nearest thousandth.

54. Solve the system. If the system has no solution, say it is inconsistent. Use either substitution or elimination.

$$\begin{cases} x - 2y + z = 0 \\ 3x + y - z = 7 \\ x - y - z = 1 \end{cases}$$

55. An economy consists of three industries, A, B, and C, each of which produces a single product. The interrelationship between the three industries is

	A	B	C	Current Customer Demand	Total Output
A	20	10	40	30	100
B	60	40	20	80	200
C	50	10	10	30	100

If the forecast demand in three years is $D_3 = \begin{bmatrix} 40 \\ 70 \\ 30 \end{bmatrix}$, what should the production be?

56. The following table shows the demand (in thousands of units) of a product at various prices (in dollars).

Price, x	3	6	8	10
Demand, y	100	80	72	51

Find the least squares line of the best fit to the above data

SOLUTIONS

1. (Section 2.1)
No, because (1, –1) is a point on $5x - y = 6$ but is not a point on $2x + y = 2$.
(i.e., $2(1) + (-1) \neq 2$)

2. (Section 2.1)
Multiply the first equation by three:
$$\begin{array}{r} 9x + 3y = 12 \\ x - 3y = -2 \\ \hline 10x = 10 \\ x = 1 \end{array}$$
Substituting $x = 1$ into either equation yields $y = 1$. Solution: (1, 1)

3. (Section 2.2)
$$\left[\begin{array}{cc|c} 3 & 1 & 4 \\ 1 & -3 & 8 \end{array}\right]$$

4. (Section 2.3)
$$\left[\begin{array}{ccc|c} 1 & -1 & 2 & 4 \\ 0 & 3 & 1 & 5 \\ 3 & -1 & 0 & 4 \end{array}\right]$$

5. (Section 2.3)
$$\left[\begin{array}{ccc|c} 2 & -3 & 1 & 0 \\ 1 & -1 & -2 & -2 \\ -5 & 1 & 0 & -4 \end{array}\right]$$

6. (Section 2.2)
$$\begin{bmatrix} 1 & 5 & -6 \\ 6 & -1 & 2 \end{bmatrix} \rightarrow \begin{bmatrix} 1 & 5 & -6 \\ 4 & -11 & 14 \end{bmatrix}$$

7. (Section 2.2)
$$\left[\begin{array}{cc|c} 1 & 2 & 5 \\ 1 & -1 & -1 \end{array}\right] \rightarrow \left[\begin{array}{cc|c} 1 & 2 & 5 \\ 0 & -3 & -6 \end{array}\right] \rightarrow \left[\begin{array}{cc|c} 1 & 2 & 5 \\ 0 & 1 & 2 \end{array}\right] \rightarrow \left[\begin{array}{cc|c} 1 & 0 & 1 \\ 0 & 1 & 2 \end{array}\right]$$
Unique solution: $x = 1, y = 2$

8. (Section 2.2)
$$\left[\begin{array}{ccc|c} 1 & 1 & 2 & 6 \\ -1 & -1 & 1 & -3 \\ 2 & 1 & 3 & 7 \end{array}\right] \rightarrow \left[\begin{array}{ccc|c} 1 & 1 & 2 & 6 \\ 0 & 0 & 3 & 3 \\ 0 & -1 & -1 & -5 \end{array}\right] \rightarrow \left[\begin{array}{ccc|c} 1 & 1 & 2 & 6 \\ 0 & -1 & -1 & -5 \\ 0 & 0 & 3 & 3 \end{array}\right] \rightarrow$$
$$\left[\begin{array}{ccc|c} 1 & 1 & 2 & 6 \\ 0 & 1 & 1 & 5 \\ 0 & 0 & 3 & 3 \end{array}\right] \rightarrow \left[\begin{array}{ccc|c} 1 & 0 & 1 & 1 \\ 0 & 1 & 1 & 5 \\ 0 & 0 & 3 & 3 \end{array}\right] \rightarrow \left[\begin{array}{ccc|c} 1 & 0 & 1 & 1 \\ 0 & 1 & 1 & 5 \\ 0 & 0 & 1 & 1 \end{array}\right] \rightarrow$$
$$\left[\begin{array}{ccc|c} 1 & 0 & 1 & 1 \\ 0 & 1 & 1 & 5 \\ 0 & 0 & 1 & 1 \end{array}\right] \rightarrow \left[\begin{array}{ccc|c} 1 & 0 & 0 & 0 \\ 0 & 1 & 0 & 4 \\ 0 & 0 & 1 & 1 \end{array}\right]$$
Unique solution: $x = 0, y = 4, z = 1$

9. (Section 2.2)

$$\left[\begin{array}{cccc|c} 1 & 2 & 0 & 10 & 17 \\ 1 & 3 & 0 & 14 & 25 \\ 2 & 2 & -1 & 10 & 14 \\ 2 & 1 & 0 & 10 & 12 \end{array}\right] \rightarrow \left[\begin{array}{cccc|c} 1 & 2 & 0 & 10 & 17 \\ 0 & 1 & 0 & 4 & 8 \\ 0 & -2 & -1 & -10 & -20 \\ 0 & -3 & 0 & -10 & -22 \end{array}\right]$$

$$\left[\begin{array}{cccc|c} 1 & 0 & 0 & 2 & 1 \\ 0 & 1 & 0 & 4 & 8 \\ 0 & 0 & -1 & -2 & -4 \\ 0 & 0 & 0 & 2 & 2 \end{array}\right] \rightarrow \left[\begin{array}{cccc|c} 1 & 0 & 0 & 2 & 1 \\ 0 & 1 & 0 & 4 & 8 \\ 0 & 0 & 1 & 2 & 4 \\ 0 & 0 & 0 & 2 & 2 \end{array}\right]$$

$$\left[\begin{array}{cccc|c} 1 & 0 & 0 & 2 & 1 \\ 0 & 1 & 0 & 4 & 8 \\ 0 & 0 & 1 & 2 & 4 \\ 0 & 0 & 0 & 1 & 1 \end{array}\right] \rightarrow \left[\begin{array}{cccc|c} 1 & 0 & 0 & 0 & -1 \\ 0 & 1 & 0 & 0 & 4 \\ 0 & 0 & 1 & 0 & 2 \\ 0 & 0 & 0 & 1 & 1 \end{array}\right]$$

Unique solution: $x_1 = -1$, $x_2 = 4$, $x_3 = 2$, $x_4 = 1$

10. (Section 2.2)

$$\left[\begin{array}{ccc|c} 1 & 1 & 1 & 3 \\ 1 & 2 & 1 & 0 \\ 1 & 3 & 1 & 1 \end{array}\right] \rightarrow \left[\begin{array}{ccc|c} 1 & 1 & 1 & 3 \\ 0 & 1 & 0 & -3 \\ 0 & 2 & 0 & -2 \end{array}\right] \rightarrow \left[\begin{array}{ccc|c} 1 & 0 & 1 & 6 \\ 0 & 1 & 0 & -3 \\ 0 & 0 & 0 & 4 \end{array}\right]$$

No solution

11. (Section 2.3)

$$\left[\begin{array}{cc|c} 1 & -1 & -2 \\ 3 & -2 & -2 \\ 5 & -3 & -2 \end{array}\right] \rightarrow \left[\begin{array}{cc|c} 1 & -1 & -2 \\ 0 & 1 & 4 \\ 0 & 2 & 8 \end{array}\right] \rightarrow \left[\begin{array}{cc|c} 1 & 0 & 2 \\ 0 & 1 & 4 \\ 0 & 0 & 0 \end{array}\right]$$

Unique solution: $x = 2$, $y = 4$

12. (Section 2.3)

$$\left[\begin{array}{ccc|c} 1 & 1 & 1 & 3 \\ 2 & -1 & -2 & -3 \\ 1 & 3 & -4 & -3 \end{array}\right] \rightarrow \left[\begin{array}{ccc|c} 1 & 1 & 1 & 3 \\ 0 & -3 & -4 & -9 \\ 0 & 3 & -4 & -3 \end{array}\right] \rightarrow \left[\begin{array}{ccc|c} 1 & 1 & 1 & 3 \\ 0 & 1 & 4/3 & 3 \\ 1 & 3 & -4 & -3 \end{array}\right]$$

$$\left[\begin{array}{ccc|c} 1 & 0 & -1/3 & 0 \\ 0 & 1 & 4/3 & 3 \\ 0 & 0 & -8 & 12 \end{array}\right] \rightarrow \left[\begin{array}{ccc|c} 1 & 0 & -1/3 & 0 \\ 0 & 1 & 4/3 & 3 \\ 0 & 0 & 1 & 3/2 \end{array}\right] \rightarrow \left[\begin{array}{ccc|c} 1 & 0 & 0 & 1/2 \\ 0 & 1 & 0 & 1 \\ 0 & 0 & 1 & 3/2 \end{array}\right]$$

Unique solution: $x_1 = 1/2$, $x_2 = 1$, $x_3 = 3/2$

13. (Section 2.1)
$x = 2$, $y = 2$

14. (Section 2.1)
$x = 4$, $y = -3$

15. (Section 2.1)
Inconsistent system

16. (Section 2.1)
$x = -2$, $y = -3$

17. (Section 2.3)

$$\left[\begin{array}{ccc|c} 1 & 2 & -1 & 2 \\ 2 & 2 & -3 & 1 \end{array}\right] \rightarrow \left[\begin{array}{ccc|c} 1 & 2 & -1 & 2 \\ 0 & -2 & -1 & -3 \end{array}\right] \rightarrow \left[\begin{array}{ccc|c} 1 & 2 & -1 & 2 \\ 0 & 1 & 1/2 & 3/2 \end{array}\right] \rightarrow$$

$$\left[\begin{array}{ccc|c} 1 & 0 & -2 & -1 \\ 0 & 1 & 1/2 & 3/2 \end{array}\right]$$

Infinitely many solutions: $x_1 = -1 + 2x_3$
$x_2 = 3/2 - (1/2)x_3$

18. (Section 2.3)

$$\left[\begin{array}{cccc|c} 0 & 0 & 1 & 1 & 7 \\ -1 & 1 & 0 & 1 & 5 \\ 1 & 1 & 1 & -1 & 6 \\ 0 & -1 & 0 & 1 & 10 \end{array}\right] \rightarrow \left[\begin{array}{cccc|c} 1 & 1 & 1 & -1 & 6 \\ -1 & 1 & 0 & 1 & 5 \\ 0 & 0 & 1 & 1 & 7 \\ 0 & -1 & 0 & 1 & 10 \end{array}\right] \rightarrow \left[\begin{array}{cccc|c} 1 & 1 & 1 & -1 & 6 \\ 0 & 2 & 1 & 0 & 11 \\ 0 & 0 & 1 & 1 & 7 \\ 0 & -1 & 0 & 1 & 10 \end{array}\right]$$

$$\left[\begin{array}{cccc|c} 1 & 1 & 1 & -1 & 6 \\ 0 & -1 & 0 & 1 & 10 \\ 0 & 0 & 1 & 1 & 7 \\ 0 & 2 & 1 & 0 & 11 \end{array}\right] \rightarrow \left[\begin{array}{cccc|c} 1 & 1 & 1 & -1 & 6 \\ 0 & 1 & 0 & -1 & -10 \\ 0 & 0 & 1 & 1 & 7 \\ 0 & 2 & 1 & 0 & 11 \end{array}\right] \rightarrow \left[\begin{array}{cccc|c} 1 & 0 & 1 & 0 & 16 \\ 0 & 1 & 0 & -1 & -10 \\ 0 & 0 & 1 & 1 & 7 \\ 0 & 0 & 1 & 2 & 31 \end{array}\right]$$

$$\left[\begin{array}{cccc|c} 1 & 0 & 0 & -1 & 9 \\ 0 & 1 & 0 & -1 & -10 \\ 0 & 0 & 1 & 1 & 7 \\ 0 & 0 & 0 & 1 & 24 \end{array}\right] \rightarrow \left[\begin{array}{cccc|c} 1 & 0 & 0 & 0 & 33 \\ 0 & 1 & 0 & 0 & 14 \\ 0 & 0 & 1 & 0 & -17 \\ 0 & 0 & 0 & 1 & 24 \end{array}\right]$$

Unique solution:
$x_1 = 33$, $x_2 = 14$
$x_3 = -17$, $x_4 = 24$

19. (Section 2.3)

$$\left[\begin{array}{ccc|c} 1 & 1 & 0 & 5 \\ 0 & 1 & -1 & 8 \\ 1 & 0 & 1 & -2 \end{array}\right] \rightarrow \left[\begin{array}{ccc|c} 1 & 1 & 0 & 5 \\ 0 & 1 & -1 & 8 \\ 1 & -1 & 1 & -7 \end{array}\right] \rightarrow \left[\begin{array}{ccc|c} 1 & 0 & 1 & -3 \\ 0 & 1 & -1 & 8 \\ 0 & 0 & 0 & 1 \end{array}\right]$$

No solution

20. (Section 2.3)

$$\left[\begin{array}{cccc|c} 1 & 0 & 2 & -1 & 5 \\ 2 & 1 & 3 & 0 & 7 \\ 5 & 1 & 9 & -3 & 12 \end{array}\right] \rightarrow \left[\begin{array}{cccc|c} 1 & 0 & 2 & -1 & 5 \\ 0 & 1 & -1 & 2 & -3 \\ 0 & 1 & -1 & 2 & 2 \end{array}\right]$$

$$\left[\begin{array}{cccc|c} 1 & 0 & 2 & -1 & 5 \\ 0 & 1 & -1 & 2 & -3 \\ 0 & 0 & 0 & 0 & 5 \end{array}\right] \rightarrow \left[\begin{array}{cccc|c} 1 & 0 & 2 & -1 & 5 \\ 0 & 1 & -1 & 2 & -3 \\ 0 & 0 & 0 & 0 & 1 \end{array}\right]$$

No solution

21. (Section 2.1)

$$\begin{cases} x - y = 12 \\ 3y = 2x - 4 \end{cases} \rightarrow \begin{cases} x - y = 12 \\ 2x - 3y = 4 \end{cases} \rightarrow \begin{cases} -2x + 2y = -24 \\ 2x - 3y = 4 \end{cases}$$

$$-y = -20 \qquad y = 20$$

The smaller number is 20.

22. (Section 2.3)
One solution

23. (Section 2.3)
Infinitely many solutions

24. (Section 2.3)
No solution

25. (Section 2.4)
4×3

26. (Section 2.4)
2×3

27. (Section 2.4)
$x = 3x + 5$ $\quad$ $2y = y - 2$
$-2x = 5$ $\quad$ $y = -2$
$x = -5/2$

28. (Section 2.4)
$x + y = 8$
$x - y = 8$
$2x = 16$
$x = 8, y = 0$

29. (Section 2.4)
$$\begin{bmatrix} 6 & -6 & -3 \\ 5 & 4 & 11 \end{bmatrix}$$

30. (Section 2.4)
$$\begin{bmatrix} 4 & -12 & 4 \\ 0 & 8 & 4 \end{bmatrix} - \begin{bmatrix} 8 & -1 & 1 \\ 2 & 0 & 1 \end{bmatrix} = \begin{bmatrix} -4 & -11 & 3 \\ -2 & 8 & 3 \end{bmatrix}$$

31. (Section 2.5)
AB is defined; its dimension is 2×5.
BA is not defined.

32. (Section 2.5)
$$\begin{bmatrix} 4 & 6 \\ 8 & 5 \end{bmatrix}$$

33. (Section 2.5)
$$\begin{bmatrix} 10 & -3 & 8 \\ 2 & -1 & 0 \end{bmatrix}$$

34. (Section 2.5)
$$\begin{bmatrix} 1 & 0 \\ 0 & 1 \end{bmatrix}$$

35. (Section 2.6)

$$\left[\begin{array}{cc|cc} 2 & 3 & 1 & 0 \\ -1 & -2 & 0 & 1 \end{array}\right] \rightarrow \left[\begin{array}{cc|cc} -1 & -2 & 0 & 1 \\ 2 & 3 & 1 & 0 \end{array}\right] \rightarrow \left[\begin{array}{cc|cc} 1 & 2 & 0 & -1 \\ 2 & 3 & 1 & 0 \end{array}\right]$$

$$\left[\begin{array}{cc|cc} 1 & 2 & 0 & -1 \\ 0 & -1 & 1 & 2 \end{array}\right] \rightarrow \left[\begin{array}{cc|cc} 1 & 2 & 0 & -1 \\ 0 & 1 & -1 & -2 \end{array}\right] \rightarrow \left[\begin{array}{cc|cc} 1 & 0 & 2 & 3 \\ 0 & 1 & -1 & -2 \end{array}\right]$$

The inverse is: $\begin{bmatrix} 2 & 3 \\ -1 & -2 \end{bmatrix}$

36. (Section 2.6)

$$\left[\begin{array}{ccc|ccc} 1 & 2 & 1 & 1 & 0 & 0 \\ 1 & 1 & 2 & 0 & 1 & 0 \\ 2 & 0 & 2 & 0 & 0 & 1 \end{array}\right] \rightarrow \left[\begin{array}{ccc|ccc} 1 & 2 & 1 & 1 & 0 & 0 \\ 0 & -1 & 1 & -1 & 1 & 0 \\ 0 & -4 & 0 & -2 & 0 & 1 \end{array}\right]$$

$$\left[\begin{array}{ccc|ccc} 1 & 2 & 1 & 1 & 0 & 0 \\ 0 & 1 & -1 & 1 & -1 & 0 \\ 0 & -4 & 0 & -2 & 0 & 1 \end{array}\right] \rightarrow \left[\begin{array}{ccc|ccc} 1 & 0 & 3 & -1 & 2 & 0 \\ 0 & 1 & -1 & 1 & -1 & 0 \\ 0 & 0 & -4 & 2 & -4 & 1 \end{array}\right]$$

$$\left[\begin{array}{ccc|ccc} 1 & 0 & 3 & -1 & 2 & 0 \\ 0 & 1 & -1 & 1 & -1 & 0 \\ 0 & 0 & 1 & -1/2 & 1 & -1/4 \end{array}\right] \rightarrow \left[\begin{array}{ccc|ccc} 1 & 0 & 0 & 1/2 & -1 & 3/4 \\ 0 & 1 & 0 & 1/2 & 0 & -1/4 \\ 0 & 0 & 1 & -1/2 & 1 & -1/4 \end{array}\right]$$

The inverse is $\begin{bmatrix} 1/2 & -1 & 3/4 \\ 1/2 & 0 & -1/4 \\ -1/2 & 1 & -1/4 \end{bmatrix}$

37. (Section 2.6)

$$\left[\begin{array}{cc|cc} 8 & -4 & 1 & 0 \\ 4 & -2 & 0 & 1 \end{array}\right] \rightarrow \left[\begin{array}{cc|cc} 1 & -1/2 & 1/8 & 0 \\ 4 & -2 & 0 & 1 \end{array}\right] \rightarrow \left[\begin{array}{cc|cc} 1 & -1/2 & 1/8 & 0 \\ 0 & 0 & -1/2 & 1 \end{array}\right]$$

Since the identity cannot be obtained on the left side, the original matrix has no inverse.

38. (Section 2.6)

$$\left[\begin{array}{ccc|ccc} 2 & 1 & 1 & 1 & 0 & 0 \\ 0 & 1 & 2 & 0 & 1 & 0 \\ 2 & 2 & 1 & 0 & 0 & 1 \end{array}\right] \rightarrow \left[\begin{array}{ccc|ccc} 1 & 1/2 & 1/2 & 1/2 & 0 & 0 \\ 0 & 1 & 2 & 0 & 1 & 0 \\ 2 & 2 & 1 & 0 & 0 & 1 \end{array}\right]$$

$$\left[\begin{array}{ccc|ccc} 1 & 1/2 & 1/2 & 1/2 & 0 & 0 \\ 0 & 1 & 2 & 0 & 1 & 0 \\ 0 & 1 & 0 & -1 & 0 & 1 \end{array}\right] \rightarrow \left[\begin{array}{ccc|ccc} 1 & 0 & 1/2 & 1 & 0 & -1/2 \\ 0 & 1 & 2 & 0 & 1 & 0 \\ 0 & 0 & -2 & -1 & -1 & 1 \end{array}\right]$$

$$\left[\begin{array}{ccc|ccc} 1 & 0 & 1/2 & 1 & 0 & -1/2 \\ 0 & 1 & 2 & 0 & 1 & 0 \\ 0 & 0 & 1 & 1/2 & 1/2 & -1/2 \end{array}\right] \rightarrow \left[\begin{array}{ccc|ccc} 1 & 0 & 0 & 3/4 & -1/4 & -1/4 \\ 0 & 1 & 0 & -1 & 0 & 1 \\ 0 & 0 & 1 & 1/2 & 1/2 & -1/2 \end{array}\right]$$

$$\begin{bmatrix} x_1 \\ x_2 \\ x_3 \end{bmatrix} = \begin{bmatrix} 3/4 & -1/4 & -1/4 \\ -1 & 0 & 1 \\ 1/2 & 1/2 & -1/2 \end{bmatrix} \begin{bmatrix} 4 \\ 0 \\ 1 \end{bmatrix} = \begin{bmatrix} 11/4 \\ -3 \\ 3/2 \end{bmatrix}$$

$x_1 = 11/4$
$x_2 = -3$
$x_3 = 3/2$

39. (Section 2.7)

x_1 = amount of 5% HCl
x_2 = amount of 15% HCl
x_3 = amount of 30% HCl

$$x_1 + x_2 + x_3 = 100$$
$$.05x_1 + .15x_2 + .3x_3 = .25(100)$$

$$\left[\begin{array}{ccc|c} 1 & 1 & 1 & 100 \\ .05 & .15 & .30 & 25 \end{array}\right] \rightarrow \left[\begin{array}{ccc|c} 1 & 1 & 1 & 100 \\ 0 & .10 & .25 & 20 \end{array}\right]$$

$$\left[\begin{array}{ccc|c} 1 & 1 & 1 & 100 \\ 0 & 1 & 2.5 & 200 \end{array}\right] \rightarrow \left[\begin{array}{ccc|c} 1 & 0 & -1.5 & -100 \\ 0 & 1 & 2.5 & 200 \end{array}\right]$$

$x_1 = -100 + 1.5x_3$; $x_2 = 200 - 2.5x_3$

$x_3 = 70$, $x_1 = 5$, $x_2 = 25$
$x_3 = 80$, $x_1 = 20$, $x_2 = 0$
$x_3 = 75$, $x_1 = 12.5$, $x_2 = 12.5$

40. (Section 2.7)

x_1 = A's wages; x_2 = B's wages; x_3 = C's wages

$$\begin{bmatrix} x_1 \\ x_2 \\ x_3 \end{bmatrix} = \begin{bmatrix} 1/3 & 1/2 & 1/3 \\ 1/3 & 1/4 & 1/3 \\ 1/3 & 1/4 & 1/3 \end{bmatrix} \begin{bmatrix} x_1 \\ x_2 \\ x_3 \end{bmatrix}$$

$$x_1 = (1/3)x_1 + (1/2)x_2 + (1/3)x_3$$
$$x_2 = (1/3)x_1 + (1/4)x_2 + (1/3)x_3$$
$$x_3 = (1/3)x_1 + (1/4)x_2 + (1/3)x_3$$

$$(-2/3)x_1 + (1/2)x_2 + (1/3)x_3 = 0$$
$$(1/3)x_1 - (3/4)x_2 + (1/3)x_3 = 0$$
$$(1/3)x_1 + (1/4)x_2 - (2/3)x_3 = 0$$

$$\left[\begin{array}{ccc|c} -2/3 & 1/2 & 1/3 & 0 \\ 1/3 & -3/4 & 1/3 & 0 \\ 1/3 & 1/4 & -2/3 & 0 \end{array}\right] \rightarrow \left[\begin{array}{ccc|c} -4 & 3 & 2 & 0 \\ 4 & -9 & 4 & 0 \\ 4 & 3 & -8 & 0 \end{array}\right] \rightarrow \left[\begin{array}{ccc|c} 1 & -3/4 & -1/2 & 0 \\ 0 & -6 & 6 & 0 \\ 0 & 6 & -6 & 0 \end{array}\right]$$

$$\left[\begin{array}{ccc|c} 1 & -3/4 & -1/2 & 0 \\ 0 & 1 & -1 & 0 \\ 0 & 6 & -6 & 0 \end{array}\right] \rightarrow \left[\begin{array}{ccc|c} 1 & 0 & -5/4 & 0 \\ 0 & 1 & -1 & 0 \\ 0 & 0 & 0 & 0 \end{array}\right]$$

$x_1 = (5/4)x_3$; $x_2 = x_3$

$x_1 = \$25,000$, $x_2 = \$20,000$, $x_3 = \$20,000$

41. (Section 2.5)

$$\begin{bmatrix} 1 & 13 & 2 \\ 0 & -1 & 0 \\ 0 & 5 & 1 \end{bmatrix} \begin{bmatrix} T \\ E \\ S \end{bmatrix} = \begin{bmatrix} 1 & 13 & 2 \\ 0 & -1 & 0 \\ 0 & 5 & 1 \end{bmatrix} \begin{bmatrix} 20 \\ 5 \\ 19 \end{bmatrix} = \begin{bmatrix} 123 \\ -5 \\ 44 \end{bmatrix}$$

$$\begin{bmatrix} 1 & 13 & 2 \\ 0 & -1 & 0 \\ 0 & 5 & 1 \end{bmatrix} \begin{bmatrix} T \\ I \\ T \end{bmatrix} = \begin{bmatrix} 1 & 13 & 2 \\ 0 & -1 & 0 \\ 0 & 5 & 1 \end{bmatrix} \begin{bmatrix} 20 \\ 9 \\ 20 \end{bmatrix} = \begin{bmatrix} 177 \\ -9 \\ 65 \end{bmatrix}$$

The encoded message is: 123 –5 44 177 –9 65

42. (Section 2.8)

$$\begin{bmatrix} 2 & 1 \\ 3 & -1 \\ 0 & 5 \end{bmatrix}$$

43. (Section 2.8)

$$A = \begin{bmatrix} 2 & 1 \\ 4 & 1 \\ 6 & 1 \\ 8 & 1 \end{bmatrix} \quad X = \begin{bmatrix} a \\ b \end{bmatrix} \quad Y = \begin{bmatrix} 8 \\ 10 \\ 11 \\ 15 \end{bmatrix} \quad A^TAX = A^TY$$

$$\begin{bmatrix} 2 & 4 & 6 & 8 \\ 1 & 1 & 1 & 1 \end{bmatrix} \begin{bmatrix} 2 & 1 \\ 4 & 1 \\ 6 & 1 \\ 8 & 1 \end{bmatrix} \begin{bmatrix} a \\ b \end{bmatrix} = \begin{bmatrix} 2 & 4 & 6 & 8 \\ 1 & 1 & 1 & 1 \end{bmatrix} \begin{bmatrix} 8 \\ 10 \\ 11 \\ 15 \end{bmatrix}$$

$$\begin{bmatrix} 120 & 20 \\ 20 & 4 \end{bmatrix} \begin{bmatrix} a \\ b \end{bmatrix} = \begin{bmatrix} 242 \\ 44 \end{bmatrix}$$

$$120a + 20b = 242$$
$$20a + 4b = 44 \qquad a = 11/10, \ b = 11/2$$

The line of best fit is $y = \frac{11}{10}x + \frac{11}{2}$

44. (Section 2.4)
2×3

45. (Section 2.4)
not possible

46. (Section 2.6)
2×2

47. (Section 2.6)
not possible

48. (Section 2.8)
2×2

49. (Section 2.8)
3×3

50. (Section 2.5)
2×3

51. (Section 2.5)
2×2

52. (Section 2.6)

$$\cong \begin{bmatrix} -.008 & .021 & -.013 \\ -.024 & .014 & -.037 \\ .005 & .002 & -.012 \end{bmatrix}$$

53. (Section 2.6)

$$\begin{bmatrix} 21 & 3 & -20 \\ 18 & -4 & 31 \\ 10 & -51 & 13 \end{bmatrix} \begin{bmatrix} x \\ y \\ z \end{bmatrix} = \begin{bmatrix} 5 \\ -7 \\ 110 \end{bmatrix}$$

$x \cong .044$
$y \cong -2.288$
$z \cong -.547$

54. (Section 2.1)

$$\begin{array}{rl} x - 2y + z &= 0 \\ 3x + y - z &= 7 \\ \hline 4x - y &= 7 \end{array} \qquad \begin{array}{rl} x - 2y + z &= 0 \\ x - y - z &= 1 \\ \hline 2x - 3y &= 1 \end{array} \qquad \begin{array}{rl} 4x - y &= 7 \\ -4x + 6y &= -2 \\ \hline 5y &= 5 \end{array} \qquad y = 1$$

$2x - 3(1) = 1$; $2x = 4$; $x = 2$

$2 - 2(1) + z = 0$; $z = 0$ Therefore, $(2, 1, 0)$

55. (Section 2.7)

$$A = \begin{bmatrix} 0.2 & 0.1 & 0.4 \\ 0.3 & 0.2 & 0.1 \\ 0.5 & 0.1 & 0.1 \end{bmatrix}$$

$$X = (I - A)^{-1}D = \begin{bmatrix} 130.5 \\ 151.8 \\ 122.8 \end{bmatrix}$$

56. (Section 2.8)

$$\begin{bmatrix} 3 & 6 & 8 & 10 \\ 1 & 1 & 1 & 1 \end{bmatrix} \begin{bmatrix} 3 & 1 \\ 6 & 1 \\ 8 & 1 \\ 10 & 1 \end{bmatrix} \begin{bmatrix} m \\ b \end{bmatrix} = \begin{bmatrix} 3 & 6 & 8 & 10 \\ 1 & 1 & 1 & 1 \end{bmatrix} \begin{bmatrix} 100 \\ 80 \\ 72 \\ 51 \end{bmatrix}$$

$m = -6.70$, $b = 120.98$

$y = -6.70x + 120.98$

CHAPTER 3

LINEAR PROGRAMMING: GEOMETRIC APPROACH

1. Graph the inequality.
$$2x + 3y \geq 6$$

2. Graph the inequality.
$$x \leq 3$$

3. Graph the system of inequalities.
$$x \geq 2, \; y \leq 4$$

4. Graph the system of inequalities.
$$x \geq 0, \; y \geq 0, \; x + y \leq 8$$

In problems 5 –10, graph each system of linear inequalities. Tell whether the graph is bounded or unbounded and list each corner point of the graph.

5. $x \geq 0, \; y \geq 0, \; 2x + 3y \geq 6$

6. $x \geq 0, \; y \geq 0, \; x + y \leq 4, \; 2x + y \leq 6$

7. $x \geq 0, \; y \geq 0, \; x + y \geq 2, \; 3x + 2y \leq 12$

8. $x \geq 0, \; y \geq 0, \; x + y \leq 10, \; 2x + y \leq 8, \; x + y \geq 2$

9. $x \geq 0, \; y \geq 0, \; x + 3y \leq 4, \; 2x + 3y \leq 5$

10. $x \geq 0, \; y \geq 0, \; 3x + y \geq 4, \; 2x + 3y \geq 5$

11. Maximize $z = 2x + 4y$ subject to the constraints
$x \geq 0, \; y \geq 0, \; x + y \geq 1, \; 3x + 2y \leq 6$

12. Maximize $z = 3x + 2y$ subject to the constraints
$x \geq 0, \; y \geq 0, \; x + 2y \leq 12, \; x + y \leq 10, \; x + y \geq 4$

13. Minimize $z = 4x + 3y$ subject to the constraints
$x \geq 0, \; y \geq 0, \; x + y \geq 1, \; 2x + 3y \leq 6$

14. Minimize $z = 2x + 5y$ subject to the constraints
$x \geq 0, \; y \geq 0, \; 2x + y \leq 12, \; x + y \leq 10, \; x + y \geq 4$

15. Maximize $z = 3x + 5y$ subject to the constraints
$x \geq 0, \; y \geq 0, \; y \leq 8, \; x + y \geq 2, \; 4x + y \leq 12$

16. Minimize $z = 3x + 5y$ subject to the constraints
$x \geq 0, \; y \geq 0, \; y \leq 8, \; x + y \geq 2, \; 4x + y \leq 12$

17. An ice skate manufacturer makes two types of ice skates: racing and figure. Use the information below to determine how many of each type should be made to achieve a maximum profit.

	Racing	Figure	Maximum Time Available
Finishing Time per Pair	2 hours	3 hours	70
Manufacturing Time per Pair	1 hour	2 hours	40
Profit per Pair	$40	$50	

18. A certain diet requires at least 60 units of protein, no more than 120 units of sodium, and no more than 80 units of fat. From each serving of food A there is derived 30 units of protein, 30 units of sodium, and 10 units of fat. From each serving of food B, there is derived 20 units of protein, 20 units of sodium, and 20 units of fat. If food A costs $1.00 per serving and food B costs $0.90 per serving, how many servings of each food can there be while keeping cost at a minimum?

19. Which of the points $P_1(2,3)$, $P_2(-1,4)$, or $P_3(5,2)$ are part of the system?

$$\begin{cases} x + y \leq 4 \\ -2x + y \geq 1 \\ 2x + 3y > 10 \end{cases}$$

20. Graph the system and tell whether it is bounded or unbounded. List any corner point.

$$\begin{cases} x + y \leq 1 \\ x \geq 2 \\ y \leq 1 \end{cases}$$

Consider the figure shown below for problems 21 – 22:

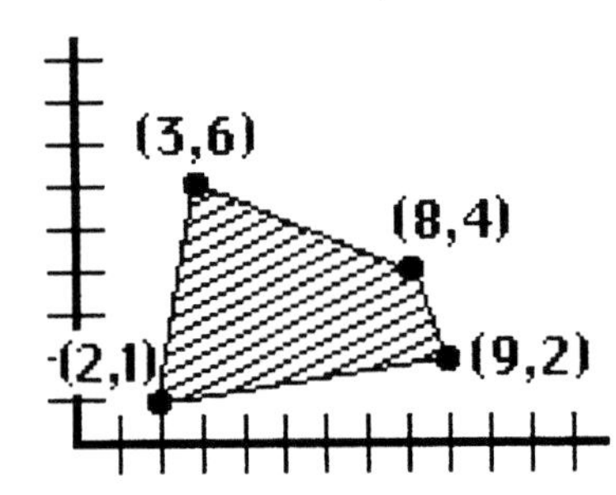

21. In a linear programming problem, the figure above represents a feasible set of points. Find the maximum and minimum of the function $z = 2x - y$.

22. Find the maximum and minimum of the function $z = x + 3y$.

23. A company makes two products, one deluxe and one regular. There are 8 hours available daily on the assembly line, and 16 hours available at the painting station. Each deluxe item takes 2 hours of assembly and 1 hour of painting. Each regular item takes 1 hour of assembly and 3 hours of painting. If the profit from each deluxe item is $420 and the profit from each regular item is $360, how many of each should be made daily to maximize profit?

24. A person has $100,000 to invest, some at 6% and some at 9%. If the strategy is to invest at least $50,000 at 6% and no more than $40,000 at 9%, what strategy maximizes the income?

SOLUTIONS

1. (Section 3.1)

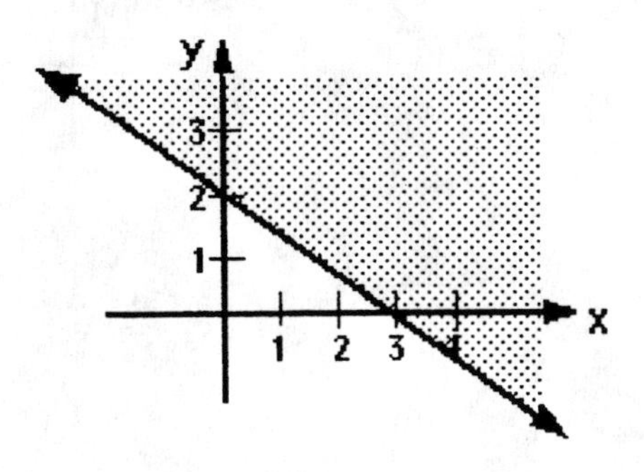

2. (Section 3.1)

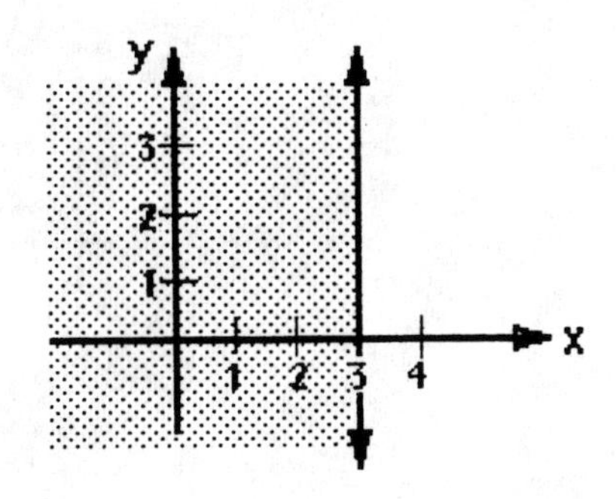

3. (Section 3.1)

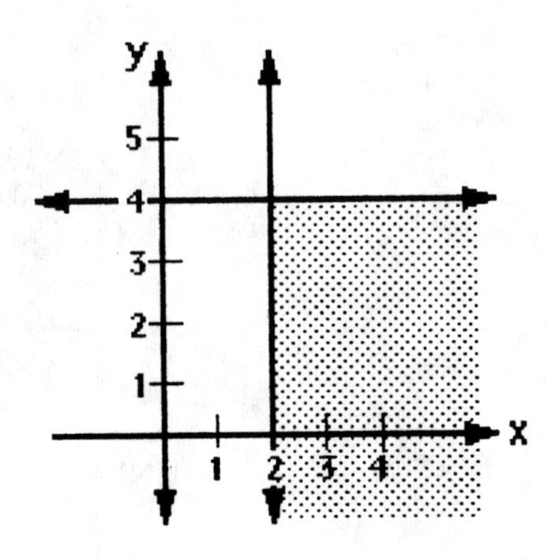

4. (Section 3.1)

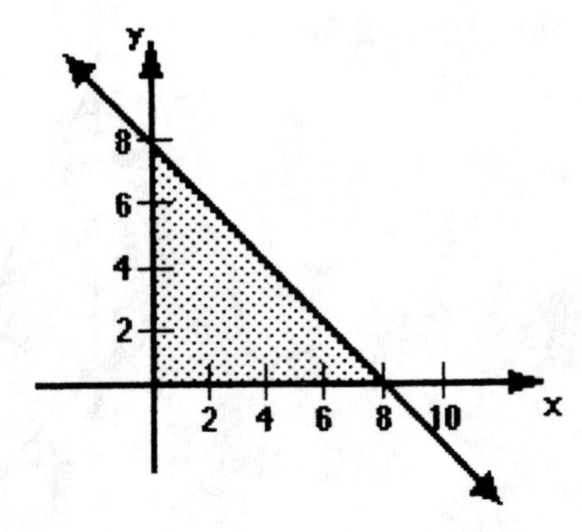

5. (Section 3.1)

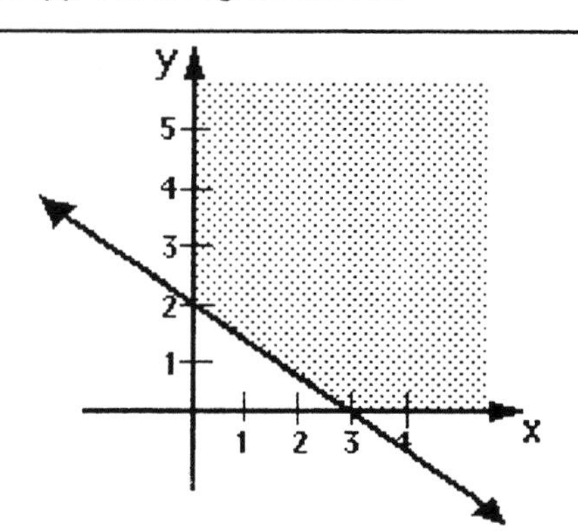

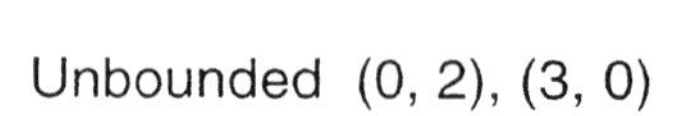
Unbounded (0, 2), (3, 0)

6. (Section 3.1)

Bounded (0, 4), (2, 2), (3, 0), (0, 0)

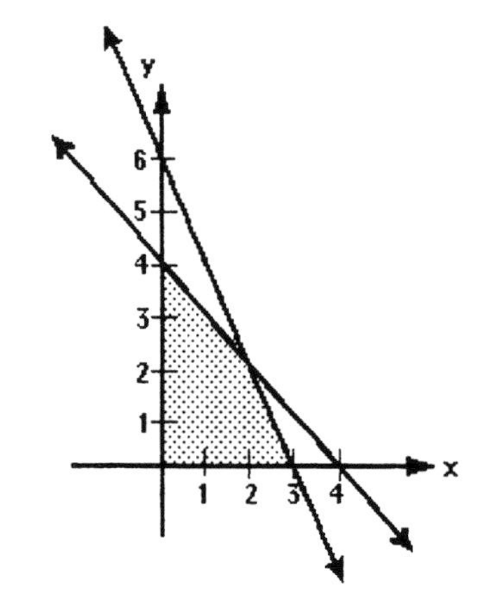

7. (Section 3.1)

Bounded (0, 6), (4, 0), (2, 0), (0, 2)

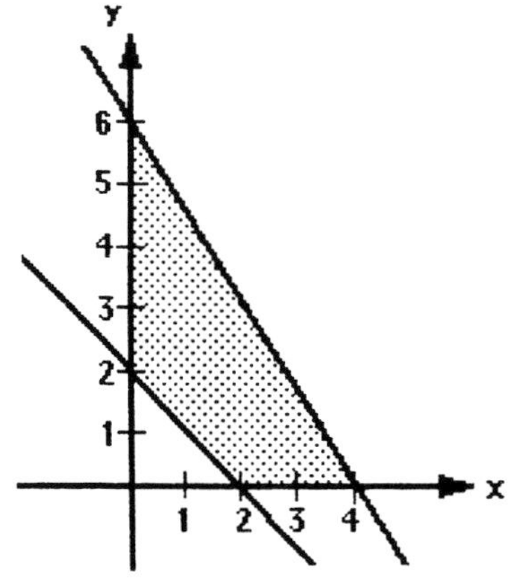

8. (Section 3.1)

Bounded (0, 2), (0, 8), (4, 0), (2, 0)

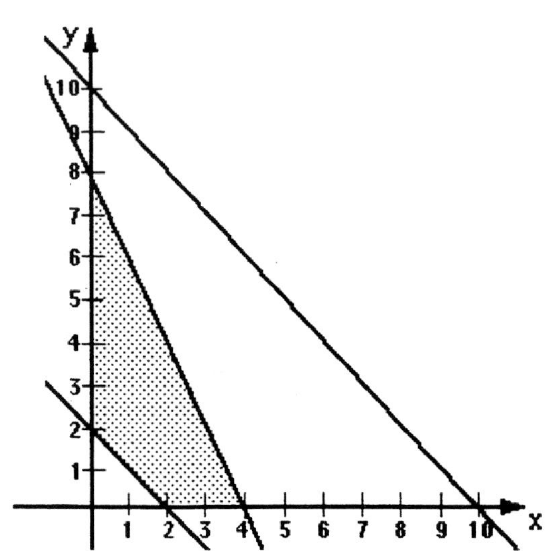

9. (Section 3.1)

Bounded (0, 0), (0, 4/3), (1, 1), (5/3, 0)

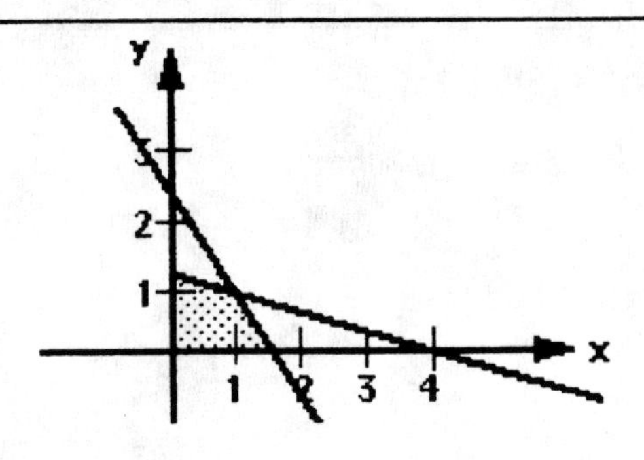

10. (Section 3.1)

Unbounded (0, 4), (1, 1), (5/2, 0)

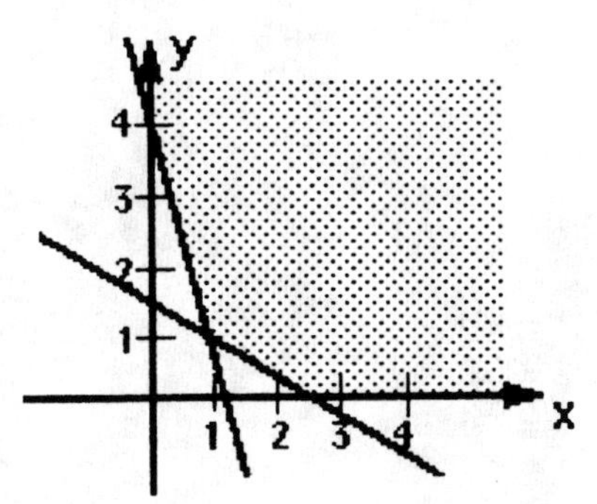

11. (Section 3.2)

(1, 0): z = 2
(2, 0): z = 4
(0, 3): z = 12 Maximum
(0, 1): z = 4

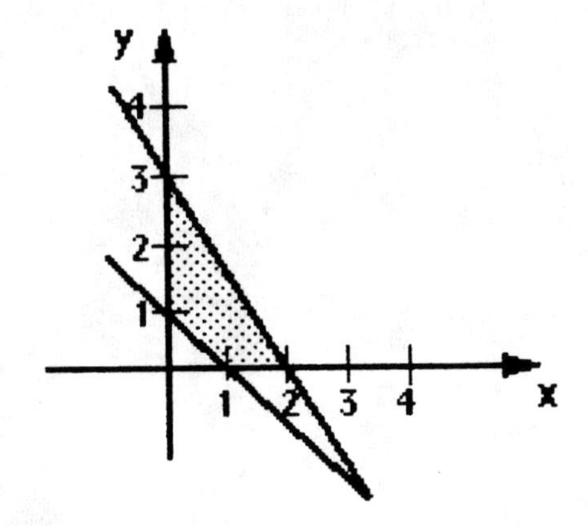

12. (Section 3.2)

(0, 4) z = 8
(0, 6) z = 12
(8, 2) z = 28
(10, 0) z = 30 Maximum
(4, 0) z = 12

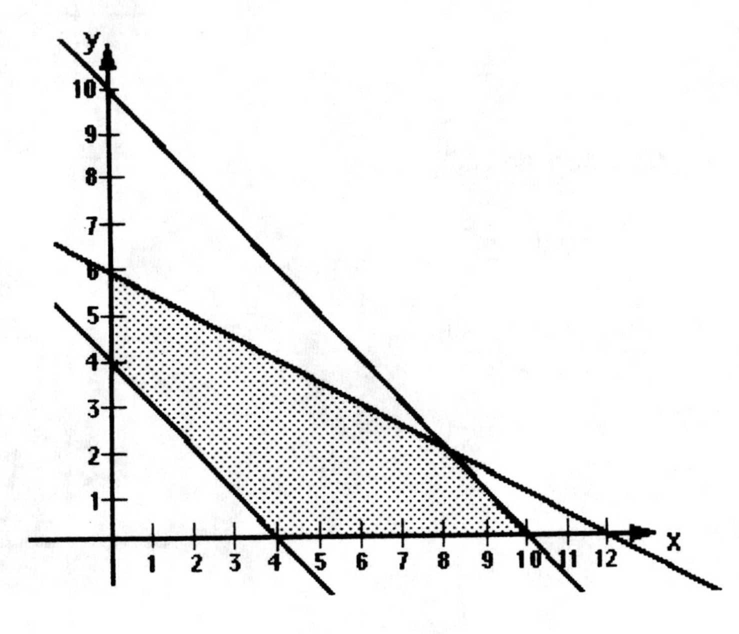

13. (Section 3.2)

(0, 1) z = 3 Minimum
(0, 2) z = 6
(1, 0) z = 4
(3, 0) z = 12

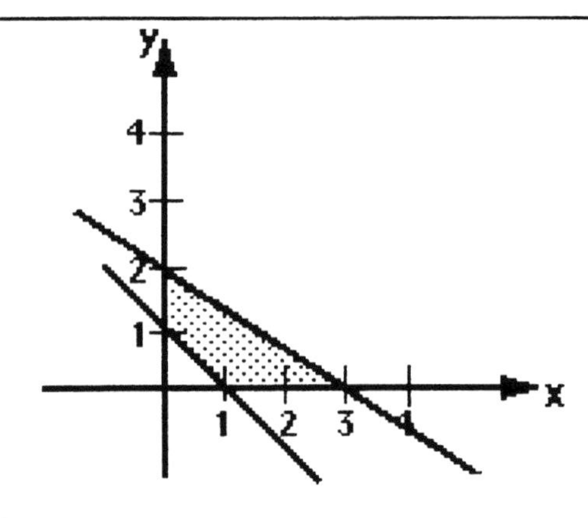

14. (Section 3.2)

(0, 4) z = 20
(0, 10) z = 50
(2, 8) z = 44
(6, 0) z = 12
(4, 0) z = 8 Minimum

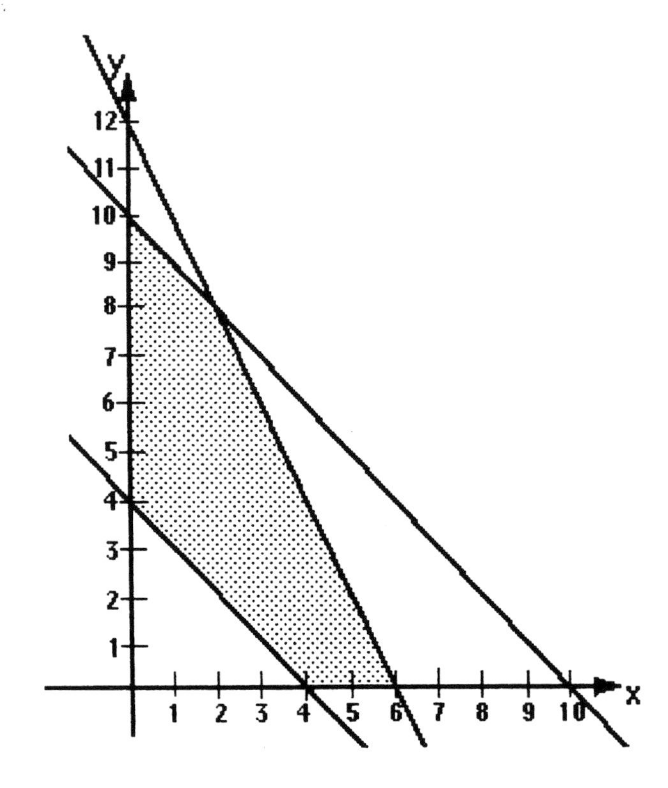

15. (Section 3.2)

(0, 8) z = 40
(1, 8) z = 43 Maximum
(3, 0) z = 9
(2, 0) z = 6
(0, 2) z = 10

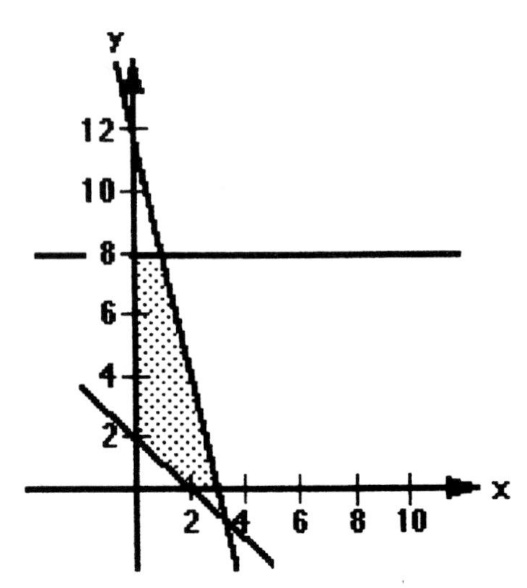

16. (Section 3.2)

(0, 8) $z = 40$
(1, 8) $z = 43$
(3, 0) $z = 9$
(2, 0) $z = 6$ Minimum
(0, 2) $z = 10$

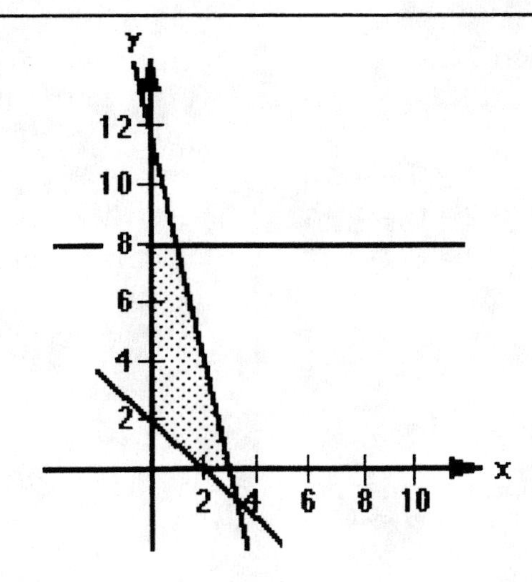

17. (Section 3.3)

x = number of racing skates
y = number of figure skates
P = profit

Maximize
$P = 40x + 50y$
subject to
$x \geq 0,\ y \geq 0,\ 2x + 3y \leq 70,\ x + 2y \leq 40$

(0, 20) P = \$1,000
(20, 10) P = \$1,300
(35, 0) P = \$1,400 Maximum

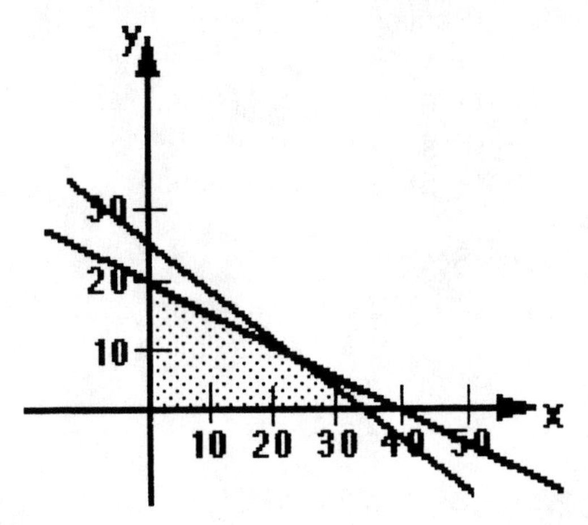

18. (Section 3.3)

	FOOD A	FOOD B
Protein	30	20
Sodium	30	20
Fat	10	20
Cost	1.00	0.90

x = servings of FOOD A
y = servings of FOOD B
C = cost

Minimize
$C = x + .9y$
subject to
$30x + 20y \geq 60$
$30x + 20y \leq 120$
$10x + 20y \leq 80$
$x \geq 0,\ y \geq 0$

(0, 3) $C = 2.70$
(0, 4) $C = 3.60$
(2, 3) $C = 4.70$
(4, 0) $C = 4.00$
(2, 0) $C = 2.00$

Use 2 servings of FOOD A to minimize cost

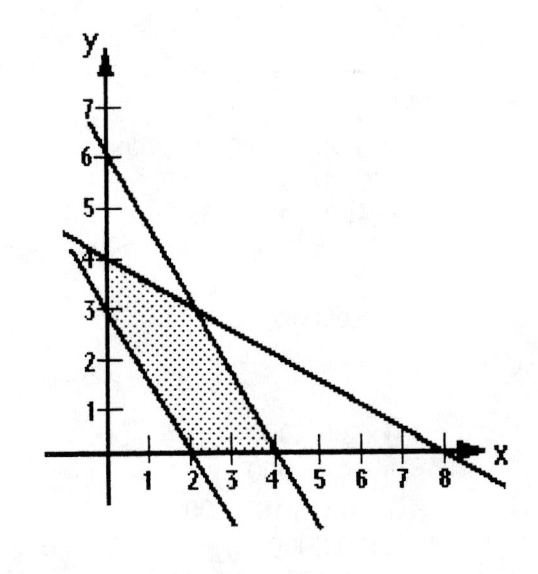

19. (Section 3.2)

P_1 $2 + 3 \neq 4$ $\therefore$ P_1 is not part of the system.

P_2
$$-1 + 4 \leq 4$$
$$-2(-1) + 4 \geq 1$$
$$2(-1) + 3(4) \geq 10$$
$\therefore$ P_2 is not part of the system.

P_3 $-2(5) + 2 \geq 1$ $\therefore$ P_3 is not part of the system.

20. (Section 3.2)

unbounded corner point (2, –1)

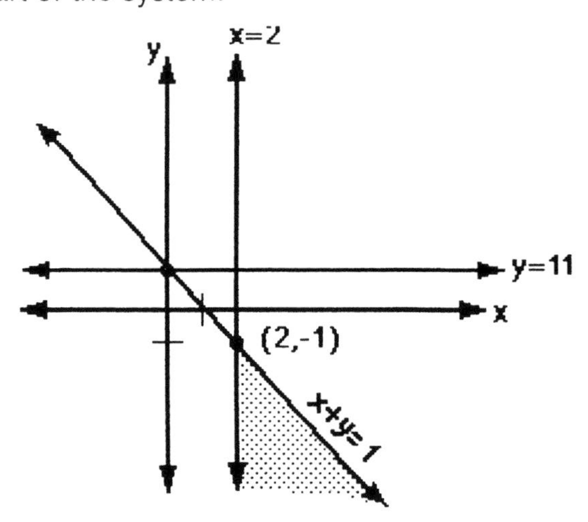

21.  (Section 3.2)

(2, 1) z = 3
(3, 6) z = 0 minimum
(8, 4) z = 12
(9, 2) z = 16 maximum

22. (Section 3.2)

(2, 1) z = 5 minimum
(3, 6) z = 21 maximum
(8, 4) z = 20
(9, 2) z = 15

23. (Section 3.3)

$z = 420x + 360y$
$x \geq 0, y \geq 0$
$2x + y \leq 8$
$2x + 3y \leq 12$
(0, 4) z = 1,440
(3, 2) z = 1,980 maximum
(0, 0) z = 0
(4, 0) z = 1,680
Make 3 deluxe items and 2 regular items,

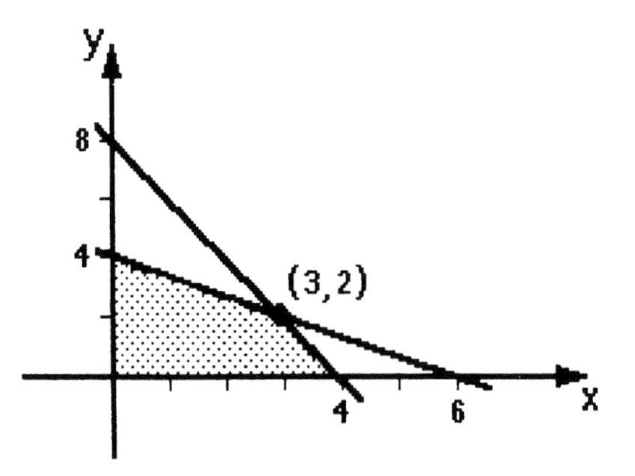

24.  (Section 3.3)

$z = 0.06x + 0.09y$
$x \geq 0, y \geq 0$
$x \geq 50{,}000$
$y \leq 40{,}000$
$x + y \leq 100{,}000$
(100,000, 0) z = 6,000
(60,000, 40,000) z = 7,200 maximum
(50,000, 0) z = 3,000
(50,000, 40,000) z = 6,600
Invest \$60,000 at 6% and \$40,000 at 9%.

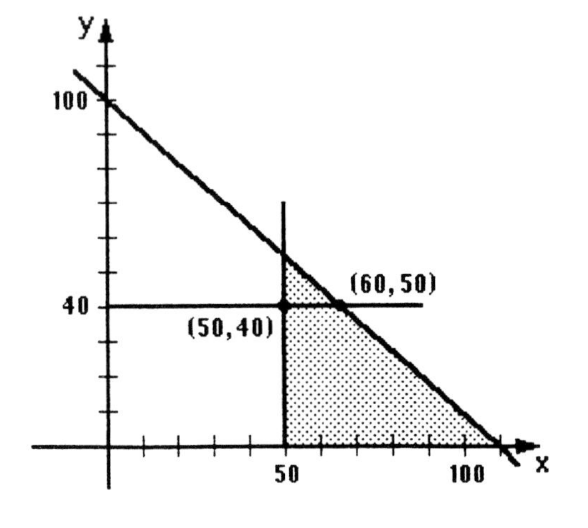

CHAPTER 4

LINEAR PROGRAMMING: SIMPLEX METHOD

In problems 1 – 4, determine if the maximum linear programming problem is in standard form. Do not solve.

1. Maximize $P = 5x_1 + 2x_2$

   subject to the constraints $2x_1 + 3x_2 \leq 6$, $x_1 + 4x_2 \leq 4$, $x_1 \geq 0, x_2 \geq 0$

2. Maximize $P = 2x_1 + 7x_2 + 8x_3$

   subject to the constraints $x_1 + x_2 + x_3 \leq 8$, $x_1 \geq 0, x_2 \geq 0$

3. Maximize $P = x_1 + 8x_2 + x_3$

   subject to the constraints $3x_1 + 4x_2 \leq 8$, $x_1 + x_2 + x_3 \geq 1$, $x_1 \geq 0, x_2 \geq 0, x_3 \geq 0$

4. Maximize $P = 2x_1 + 3x_2 + 8x_3$

   subject to the constraints $0 \leq x_1 \leq 4$, $x_2 + x_3 + x_4 \leq 8$, $x_2 \geq 0, x_3 \geq 0, x_4 \geq 0$

In problems 5 – 8, each maximum problem is not in standard form. Determine if the problem can be modified so as to be in standard form. If it can, write the modified version.

5. Maximize $P = 3x_1 + 5x_2$

   subject to the constraints $2x_1 + 3x_2 \leq 5$, $x_1 \geq 0$

6. Maximize $P = 5x_1 + 3x_2$

   subject to the constraints $4x_1 - 2x_2 \geq -2$, $x_1 + x_2 \leq 5$, $x_1 \geq 0, x_2 \geq 0$

7. Maximize $P = x_1 + 2x_2 + 3x_3$

   subject to the constraints $x_1 + x_2 + x_3 \leq 8$, $x_1 + x_2 \geq -3$, $x_1 \geq 0, x_2 \geq 0, x_3 \geq 0$

8. Maximize $P = x_1 + 2x_2 + 5x_3$

   subject to the constraints $x_1 + x_2 + x_3 \leq 9$, $2x_1 - 3x_2 + x_3 \geq 4$, $x_1 \geq 0, x_2 \geq 0, x_3 \geq 0$

In problems 9 – 10, each maximum problem is in standard form. Introduce slack variables and set up the initial simplex.

9. Maximize $P = 8x_1 + 2x_2 + 3x_3$

subject to the constraints $x_1 + 3x_2 + 2x_3 \le 10$ $\quad x_1 \ge 0, x_2 \ge 0, x_3 \ge 0$

$4x_1 + 2x_2 + 3x_3 \le 8$

10. Maximize $P = 2x_1 + 4x_2 + x_3$

subject to the constraints $x_1 + x_2 + x_3 \le 12$ $\quad x_1 \ge 0, x_2 \ge 0, x_3 \ge 0$

$x_1 + x_2 \le 6$

$2x_1 - x_2 + 3x_3 \le 6$

In problems 11 – 13, determine if the tableaux:

(a) is the final tableaux

(b) requires additional pivoting

(c) indicates no solution to the problem

If the answer is (a), write down the solution; if the answer is (b), identify the pivot element.

11.

x_1	x_2	s_1	s_2		
1	1/2	1/4	0	30	x_1
0	1	−1/2	1	20	s_2
0	−1	1	0	120	

12.

x_1	x_2	s_1	s_2		
−4	1	0	2	18	x_2
−6	0	1	4	12	s_1
−5	0	0	6	10	

13.

x_1	x_2	s_1	s_2		
6	0	1	4	10	s_1
2	1	0	8	5	x_2
5	0	0	10	25	

14. Maximize $P = 3x_1 + 4x_2$

subject to the constraints $2x_1 + 3x_2 \le 5$ $\quad x_1 \ge 0, x_2 \ge 0$

$3x_1 + 2x_2 \le 5$

15. Maximize $P = x_1 + 2x_2 + 3x_3$

subject to the constraints $2x_1 + x_2 + x_3 \le 25$ $\quad x_1 \ge 0, x_2 \ge 0, x_3 \ge 0$

$2x_1 + 3x_2 + 3x_3 \le 30$

16. Maximize $P = x_1 + 8x_2 + 6x_3$

subject to the constraints
$$5x_1 + 2x_2 + 6x_3 \le 30$$
$$2x_1 + 3x_2 + x_3 \le 9$$
$$3x_1 + 5x_2 + 3x_3 \le 20$$
$x_1 \ge 0, x_2 \ge 0$

In problems 17 – 19, determine if the given Minimum problem is in standard form. Do *not* solve.

17. Minimize $C = 5x_1 + 7x_2$

subject to the constraints
$$2x_1 - x_2 \ge 3$$
$$x_1 + x_2 \ge 5$$
$x_1 \ge 0, x_2 \ge 0$

18. Minimize $C = x_1 + x_2 + x_3$

subject to the constraints
$$2x_1 + x_2 + 3x_3 \ge -1$$
$$x_1 + 2x_2 + x_3 \le 0$$
$x_1 \ge 0, x_2 \ge 0, x_3 \ge 0$

19. Minimize $C = x_1 + 2x_2 - x_3$

subject to the constraints
$$2x_1 + x_2 + 3x_3 \ge 5$$
$$x_1 - x_2 + x_3 \ge 2$$
$x_1 \ge 0, x_2 \ge 0, x_3 \ge 0$

In problems 20 – 22, write the dual problem for each minimum linear programming problem.

20. Minimize $C = 3x_1 + 5x_3$

subject to the constraints
$$x_1 + x_2 \ge 2$$
$$3x_1 + 2x_2 \ge 6$$
$x_1 \ge 0, x_2 \ge 0$

21. Minimize $C = 4x_1 + 2x_2 + 6x_3$

subject to the constraints
$$5x_1 + 3x_2 + 3x_3 \ge 20$$
$$3x_1 + x_2 + 2x_3 \ge 9$$
$$2x_1 + 6x_2 + 5x_3 \ge 30$$
$x_1 \ge 0, x_2 \ge 0, x_3 \ge 0$

22. Minimize $C = x_1 + x_2 + 4x_3 + 2x_4$

subject to the constraints
$$x_1 + x_2 + x_3 + x_4 \le 100$$
$$x_1 + x_3 \ge 30$$
$$x_2 + x_4 \ge 50$$
$x_1 \ge 0, x_2 \ge 0, x_3 \ge 0, x_4 \ge 0$

23. Minimize $C = 4x_1 + 3x_2 + 2x_3$

subject to
$$-3x_1 - 2x_2 + x_3 \ge -2$$
$$x_1 + x_2 + x_3 \ge 2$$
$$x_1 + 2x_3 \ge 3$$
$x_1 \ge 0, x_2 \ge 0, x_3 \ge 0$

24. Minimize $C = 2x_1 + x_2 + x_3$

subject to
$$-3x_1 + x_2 + 4x_3 \geq 12$$
$$x_1 + 3x_2 + 2x_3 \geq 10$$
$$-x_1 + x_2 - x_3 \geq -8$$
$x_1 \geq 0, x_2 \geq 0, x_3 \geq 0$

25. Maximize $P = -x_1 + 2x_2 + 3x_3$

subject to
$$x_1 + 3x_2 + x_3 \leq 9$$
$$-x_1 + 3x_2 + 2x_3 \geq 2$$
$$x_1 - 2x_2 + 3x_3 \geq 5$$
$x_1 \geq 0, x_2 \geq 0, x_3 \geq 0$

26. Maximize $P = x_1 + 8x_2 + 6x_3$

subject to
$$x_1 + x_2 + x_3 \leq 10$$
$$5x_1 + 2x_2 + 6x_3 \geq 30$$
$$2x_1 + 3x_2 + x_3 \geq 9$$
$$3x_1 + 5x_2 + 3x_3 \geq 20$$
$x_1 \geq 0, x_2 \geq 0, x_3 \geq 0$

27. Minimize $C = 5x_1 + 7x_2 + 6x_3$

subject to the constraints
$$x_1 + x_2 + x_3 = 10$$
$$2x_1 + x_2 + 3x_3 \leq 19$$
$$2x_2 + 3x_3 \geq 21$$

28. A company makes three products, A, B, and C. There are 500 pounds of raw material available. Each unit of product A requires 2 pounds of raw material, each unit of product B requires 2 pounds, and each unit of product C requires 3 pounds. The assembly line has 1,000 hours of operation available. Each unit of product A requires 4 hours, while each unit of products B and C requires 5 hours. The company realizes a profit of \$500 for each unit of product A, \$600 for each unit of product B, and \$1,000 for each unit of product C. How many units of each of the three products should the company make to maximize profits?

29. A dietitian is attempting to prepare a meal from three foods, x_1, x_2, and x_3. Each food contains vitamin A, protein, and vitamin C. One unit of food x_1 contains 3 mg of vitamin A, 20 g of protein, and 3 mg of vitamin C. One unit of food x_2 contains 2 mg of vitamin A, 10 g of protein, and 3 mg of vitamin C. One unit of food x_3 contains 5 mg of vitamin A, 10 g of protein, and 3 mg of vitamin C. The diet must contain at least 10 mg of vitamin A, 40 g of protein, and 20 mg of vitamin C. Foods x_1, x_2, and x_3 contain 5 g, 2 g, and 1 g of fat, respectively. If the intent is to minimize the fat, write the dual for the linear program.

SOLUTIONS

1. (Section 4.1)
   In standard form

2. (Section 4.1)
   Not in standard form

3. (Section 4.1)
   Not in standard form

4. (Section 4.1)
   In standard form

5. (Section 4.1)
   Cannot be modified

6. (Section 4.1)
   Maximize $P = 5x_1 + 3x_2$

   subject to the constraints $-4x_1 + 2x_2 \le 2$, $x_1 + x_2 \le 5$, $x_1 \ge 0, x_2 \ge 0$

7. (Section 4.1)
   Maximize $P = x_1 + 2x_2 + 3x_3$

   subject to the constraints $x_1 + x_2 + x_3 \le 8$, $-x_1 - x_2 \le 3$, $x_1 \ge 0, x_2 \ge 0, x_3 \ge 0$

8. (Section 4.1)
   Cannot be modified

9. (Section 4.1)

BV	P	x_1	x_2	x_3	s_1	s_2	RHS
s_1	0	1	3	2	1	0	10
s_2	0	4	2	3	0	1	8
P	1	–8	–2	–3	0	0	0

10. (Section 4.1)

BV	P	x_1	x_2	x_3	s_1	s_2	s_3	RHS
s_1	0	1	1	1	1	0	0	12
s_2	0	1	1	0	0	1	0	6
s_3	0	2	–1	3	0	0	1	6
P	1	–2	–4	–1	0	0	0	0

11. (Section 4.2)
    (b) row 2, column 3 contains the pivot element

12. (Section 4.2)
(c) no solution

13. (Section 4.2)
(a) final tableaux; $P = 25$, $x_1 = 10$, $x_2 = 5$

14. (Section 4.2)
The initial tableaux is

	P	x_1	x_2	s_1	s_2	RHS
s_1	0	2	(3)	1	0	5
s_2	0	3	2	0	1	5
P	1	–3	–4	0	0	0

	P	x_1	x_2	s_1	s_2	RHS
x_2	0	2/3	1	1/3	0	5/3
s_2	0	(5/3)	0	–2/3	1	5/3
P	1	–1/3	0	4/3	0	20/3

	P	x_1	x_2	s_1	s_2	RHS
x_2	0	0	1	3/5	–2/5	1
x_1	0	1	0	–2/5	3/5	1
P	1	0	0	6/5	1/5	7

This is the final tableaux; $P = 7$, $x_1 = 1$, $x_2 = 1$

15. (Section 4.2)
The initial simplex tableaux is

	P	x_1	x_2	x_3	s_1	s_2	RHS
s_1	0	2	1	1	1	0	25
s_2	0	2	3	(3)	0	1	30
P	1	–1	–2	–3	0	0	0

	P	x_1	x_2	x_3	s_1	s_2	RHS
s_1	0	4/3	0	0	1	–1/3	15
x_3	0	2/3	1	1	0	1/3	10
P	1	1	1	0	0	1	30

This is the final tableaux; $P = 30$; $x_1 = 0$, $x_2 = 0$, $x_3 = 10$

16. (Section 4.2)
The initial simplex tableaux is

	P	x_1	x_2	x_3	s_1	s_2	s_3	RHS
s_1	0	5	2	6	1	0	0	30
s_2	0	2	(3)	1	0	1	0	9
s_3	0	3	5	3	0	0	1	20
P	1	–1	–8	–6	0	0	0	0

	P	x_1	x_2	x_3	s_1	s_2	s_3	RHS
s_1	0	11/3	0	16/3	1	–2/3	0	24
x_2	0	2/3	1	1/3	0	1/3	0	3
s_3	0	–1/3	0	(4/3)	0	–5/3	1	5
P	1	13/3	0	–10/3	0	8/3	0	24

	P	x_1	x_2	x_3	s_1	s_2	s_3	RHS
s_1	0	5	0	0	1	(6)	–4	4
x_2	0	3/4	1	0	0	3/4	–1/4	7/4
x_3	0	–1/4	0	1	0	–5/4	3/4	15/4
P	1	7/2	0	0	0	–3/2	5/2	73/2

	P	x_1	x_2	x_3	s_1	s_2	s_3	RHS
s_2	0	5/6	0	0	1/6	1	–2/3	2/3
x_2	0	1/8	1	0	–1/8	0	1/4	5/4
x_3	0	19/24	0	1	5/24	0	–1/12	55/12
P	1	19/4	0	0	1/4	0	3/2	75/2

This is the final tableaux;
$P = 75/2$; $x_1 = 0$, $x_2 = 5/4$, $x_3 = 55/12$

17. (Section 4.3)
In standard form

18. (Section 4.3)
Not in standard form

19. (Section 4.3)
Not in standard form

20. (Section 4.3)
The special matrix is

$$\left[\begin{array}{cc|c} 1 & 1 & 2 \\ 3 & 2 & 6 \\ 3 & 5 & 0 \end{array}\right]$$

20. (cont.)

Interchanging rows and columns,

$$\left[\begin{array}{cc|c} 1 & 3 & 3 \\ 1 & 2 & 5 \\ 2 & 6 & 0 \end{array}\right]$$

The dual problem is:

Maximize $P = 2y_1 + 6y_2$

subject to $y_1 + 3y_2 \leq 3$ $\quad y_1 \geq 0$

$y_1 + 2y_2 \leq 5$ $\quad y_2 \geq 0$

21. (Section 4.3)

The special matrix is

$$\left[\begin{array}{ccc|c} 5 & 3 & 3 & 20 \\ 3 & 1 & 2 & 9 \\ 2 & 6 & 5 & 30 \\ 4 & 2 & 6 & 0 \end{array}\right]$$

Interchanging rows and columns,

$$\left[\begin{array}{ccc|c} 5 & 3 & 2 & 4 \\ 3 & 1 & 6 & 2 \\ 3 & 2 & 5 & 6 \\ 20 & 9 & 30 & 0 \end{array}\right]$$

The dual problem is:

Maximize $P = 20y_1 + 9y_2 + 30y_3$

subject to $5y_1 + 3y_2 + 2y_3 \leq 4$ $\quad y_1 \geq 0, y_2 \geq 0, y_3 \geq 0$

$3y_1 + y_2 + 6y_3 \leq 2$

$3y_1 + 2y_2 + 5y_3 \leq 6$

22. (Section 4.4)

The standard form of the problem is

Minimize $C = x_1 + x_2 + 4x_3 + 2x_4$

subject to $-x_1 - x_2 - x_3 - x_4 \geq -100$ $\quad x_1 \geq 0, x_2 \geq 0, x_3 \geq 0, x_4 \geq 0$

$x_1 + x_3 \geq 30$

$x_2 + x_4 \geq 50$

The special matrix is

$$\left[\begin{array}{cccc|c} -1 & -1 & -1 & -1 & -100 \\ 1 & 0 & 1 & 0 & 30 \\ 0 & 1 & 0 & 1 & 50 \\ 1 & 1 & 4 & 2 & 0 \end{array}\right]$$

22. (cont.)

Interchanging rows and columns,

$$\left[\begin{array}{ccc|c} -1 & 1 & 0 & 1 \\ -1 & 0 & 1 & 1 \\ -1 & 1 & 0 & 4 \\ -1 & 0 & 1 & 2 \\ -100 & 30 & 50 & 0 \end{array}\right]$$

The dual problem is

Maximize $P = -100y_1 + 30y_2 + 50y_3$

subject to

$-y_1 + y_2 \le 1$ $\quad y_1 \ge 0$

$-y_1 + y_3 \le 1$ $\quad y_2 \ge 0$

$-y_1 + y_2 \le 4$ $\quad y_3 \ge 0$

$-y_1 + y_3 \le 2$ $\quad y_4 \ge 0$

23. (Section 4.3)

The special matrix is

$$\left[\begin{array}{ccc|c} -3 & -2 & 1 & -2 \\ 1 & 1 & 1 & 2 \\ 1 & 0 & 2 & 3 \\ 4 & 3 & 2 & 0 \end{array}\right]$$

Interchanging rows and columns,

$$\left[\begin{array}{ccc|c} -3 & 1 & 1 & 4 \\ -2 & 1 & 0 & 3 \\ 1 & 1 & 2 & 2 \\ -2 & 2 & 3 & 0 \end{array}\right]$$

The dual problem is

Maximize $P = -2y_1 + 2y_2 + 3y_3$

subject to

$-3y_1 + y_2 + y_3 \le 4$ $\quad y_1 \ge 0,\ y_2 \ge 0,\ y_3 \ge 0$

$-2y_1 + y_2 \le 3$

$y_1 + y_2 + 2y_3 \le 2$

The initial simplex tableaux is

P	y_1	y_2	y_3	s_1	s_2	s_3	RHS
0	−3	1	1	1	0	0	4
0	−2	1	0	0	1	0	3
1	1	1	2	0	0	1	2
0	2	−2	−3	0	0	0	0

23. (cont.)

$$\left[\begin{array}{ccccccc|c} 0 & -7/2 & 1/2 & 0 & 1 & 0 & -1/2 & 3 \\ 0 & -2 & 1 & 0 & 0 & 1 & 0 & 3 \\ 0 & 1/2 & (1/2) & 1 & 0 & 0 & 1/2 & 1 \\ \hline 1 & 7/2 & -1/2 & 0 & 0 & 0 & 3/2 & 3 \end{array}\right]$$

$$\left[\begin{array}{ccccccc|c} 0 & -4 & 0 & -1 & 1 & 0 & -1 & 2 \\ 0 & -3 & 0 & -2 & 0 & 1 & -1 & 1 \\ 0 & 1 & 1 & 2 & 0 & 0 & 1 & 2 \\ \hline 1 & 4 & 0 & 1 & 0 & 0 & 2 & 4 \end{array}\right] \begin{array}{c} s_1 \\ s_2 \\ y_2 \\ \\ \end{array}$$

This is a final tableaux; the solution to the minimum problem is
$C = 4$, $x_1 = 0$, $x_2 = 0$, $x_3 = 2$

24. (Section 4.3)
The special matrix is

$$\left[\begin{array}{cccc|c} 0 & -3 & 1 & 4 & 12 \\ 0 & 1 & 3 & 2 & 10 \\ 0 & -1 & 1 & -1 & -8 \\ \hline 1 & 2 & 1 & 1 & 0 \end{array}\right]$$

Interchanging rows and columns,

$$\left[\begin{array}{cccc|c} 0 & -3 & 1 & -1 & 2 \\ 0 & 1 & 3 & 1 & 1 \\ 0 & 4 & 2 & -1 & 1 \\ \hline 1 & 12 & 10 & -8 & 0 \end{array}\right]$$

The dual problem is
Maximize $P = 12y_1 + 10y_2 - 8y_3$

subject to
$$-3y_1 + y_2 - y_3 \le 2$$
$$y_1 + 3y_2 + y_3 \le 1$$
$$4y_1 + 2y_2 - y_3 \le 1$$
$y_1 \ge 0,\ y_2 \ge 0,\ y_3 \ge 0$

The initial simplex tableaux is

$$\left[\begin{array}{cccccc|c} y_1 & y_2 & y_3 & s_1 & s_2 & s_3 & \\ -3 & 1 & -1 & 1 & 0 & 0 & 2 \\ 1 & 3 & 1 & 0 & 1 & 0 & 1 \\ (4) & 2 & -1 & 0 & 0 & 1 & 1 \\ \hline -12 & -10 & 8 & 0 & 0 & 0 & 0 \end{array}\right]$$

24. (cont.)

$$\left[\begin{array}{cccccc|c} 0 & 5/2 & -7/4 & 1 & 0 & 3/4 & 11/4 \\ 0 & (5/2) & 5/4 & 0 & 1 & -1/4 & 3/4 \\ 1 & 1/2 & -1/4 & 0 & 0 & 1/4 & 1/4 \\ \hdashline 0 & -4 & 5 & 0 & 0 & 3 & 3 \end{array}\right]$$

$$\left[\begin{array}{cccccc|c} 0 & 0 & -3 & 1 & -1 & 1 & 2 \\ 0 & 1 & 1/2 & 0 & 2/5 & -1/10 & 3/10 \\ 1 & 0 & -1/2 & 0 & -1/5 & 3/10 & 1/10 \\ \hdashline 0 & 0 & 7 & 0 & 8/5 & 13/5 & 21/5 \end{array}\right]$$

This is a final tableaux; the solution to the minimum problem is
$C = 21/5,\ x_1 = 0,\ x_2 = 8/5,\ x_3 = 13/5$

25. (Section 4.4)
We write the constraints as

$$\begin{aligned} x_1 + 3x_2 + x_3 &\le 9 \\ x_1 - 3x_2 - 2x_3 &\le -2 \\ -x_1 + 2x_2 - 3x_3 &\le -5 \end{aligned}$$

Then

$$\begin{aligned} x_1 + 3x_2 + x_3 + s_1 &= 9 \\ x_1 - 3x_2 - 2x_3 + s_2 &= -2 \\ -x_1 + 2x_2 - 3x_3 + s_3 &= -5 \end{aligned} \qquad s_1 \ge 0,\ s_2 \ge 0,\ s_3 \ge 0$$

The initial simplex tableaux is

$$\left[\begin{array}{ccccccc|c} P & x_1 & x_2 & x_3 & s_1 & s_2 & s_3 & \\ 0 & 1 & 3 & 1 & 1 & 0 & 0 & 9 \\ 0 & 1 & -3 & -2 & 0 & 1 & 0 & -2 \\ 0 & -1 & 2 & (-3) & 0 & 0 & 1 & -5 \\ \hdashline 1 & 1 & -2 & -3 & 0 & 0 & 0 & 0 \end{array}\right]$$

Phase I applies

$$\left[\begin{array}{ccccccc|c} 0 & 2/3 & (11/3) & 0 & 1 & 0 & 1/3 & 22/3 \\ 0 & 5/3 & -13/3 & 0 & 0 & 1 & -2/3 & 4/3 \\ 0 & 1/3 & -2/3 & 1 & 0 & 0 & -1/3 & 5/3 \\ \hdashline 1 & 2 & -4 & 0 & 0 & 0 & -1 & 5 \end{array}\right]$$

Phase II applies

$$\left[\begin{array}{ccccccc|c} 0 & 2/11 & 1 & 0 & 3/11 & 0 & (1/11) & 2 \\ 0 & 27/11 & 0 & 0 & 13/11 & 1 & -3/11 & 10 \\ 0 & 5/11 & 0 & 1 & 2/11 & 0 & -3/11 & 3 \\ \hdashline 1 & 30/11 & 0 & 0 & 12/11 & 0 & -7/11 & 13 \end{array}\right]$$

25. (cont.)

Phase II applies

$$\left[\begin{array}{ccccccc|c}
0 & 2 & 11 & 0 & 3 & 0 & 1 & 22 \\
0 & 3 & 3 & 0 & 2 & 1 & 0 & 16 \\
0 & 1 & 3 & 1 & 1 & 0 & 0 & 9 \\
\hline
1 & 4 & 7 & 0 & 3 & 0 & 0 & 27
\end{array}\right]$$

The maximum value of P is 27 and it is achieved when
$x_1 = 0,\ x_2 = 0,\ x_3 = 9,\ s_1 = 0,\ s_2 = 16,\ s_3 = 22$

26. (Section 4.4)

We write the constraints as

$$\begin{aligned}
x_1 + x_2 + x_3 &\le 10 \\
-5x_1 - 2x_2 - 6x_3 &\le -30 \\
-2x_1 - 3x_2 - x_3 &\le -9 \\
-3x_1 - 5x_2 - 3x_3 &\le -20
\end{aligned}$$

Then

$$\begin{aligned}
x_1 + x_2 + x_3 + s_1 &= 10 \\
-5x_1 - 2x_2 - 6x_3 + s_2 &= -30 \\
-2x_1 - 3x_2 - x_3 + s_3 &= -9 \\
-3x_1 - 5x_2 - 3x_3 + s_4 &= -20
\end{aligned}
\qquad s_1 \ge 0,\ s_2 \ge 0,\ s_3 \ge 0,\ s_4 \ge 0$$

The initial simplex tableaux is

$$\left[\begin{array}{cccccccc|c}
P & x_1 & x_2 & x_3 & s_1 & s_2 & s_3 & s_4 & \\
0 & 1 & 1 & 1 & 1 & 0 & 0 & 0 & 10 \\
0 & -5 & -2 & (-6) & 0 & 1 & 0 & 0 & -30 \\
0 & -2 & -3 & -1 & 0 & 0 & 1 & 0 & -9 \\
0 & -3 & -5 & -3 & 0 & 0 & 0 & 1 & -20 \\
\hline
1 & -1 & -8 & -6 & 0 & 0 & 0 & 0 & 0
\end{array}\right]$$

Phase I applies

$$\left[\begin{array}{cccccccc|c}
0 & 1/6 & 2/3 & 0 & 1 & 1/6 & 0 & 0 & 5 \\
0 & 5/6 & 1/3 & 1 & 0 & -1/6 & 0 & 0 & 5 \\
0 & -7/6 & -8/3 & 0 & 0 & -1/6 & 1 & 0 & -4 \\
0 & -1/2 & (-4) & 0 & 0 & -1/2 & 0 & 1 & -5 \\
\hline
1 & 4 & -6 & 0 & 0 & -1 & 0 & 0 & 30
\end{array}\right]$$

Phase I applies

$$\left[\begin{array}{cccccccc|c}
0 & 1/12 & 0 & 0 & 1 & 1/12 & 0 & 1/6 & 25/6 \\
0 & 19/24 & 0 & 1 & 0 & -5/24 & 0 & 1/12 & 55/12 \\
0 & (-5/6) & 0 & 0 & 0 & 1/6 & 1 & -2/3 & -2/3 \\
0 & 1/8 & 1 & 0 & 0 & 1/8 & 0 & -1/4 & 5/4 \\
\hline
1 & 19/4 & 0 & 0 & 0 & -1/4 & 0 & -3/2 & 75/2
\end{array}\right]$$

26. (cont.)

Phase I applies

$$\left[\begin{array}{ccccccccc|c}
0 & 0 & 0 & 0 & 1 & 1/10 & 1/10 & 1/10 & 41/10 \\
0 & 0 & 0 & 1 & 0 & 1/20 & 19/20 & -11/20 & 79/20 \\
0 & 1 & 0 & 0 & 0 & -1/5 & -6/5 & (4/5) & 4/5 \\
0 & 0 & 1 & 0 & 0 & 3/20 & 3/20 & -7/20 & 23/20 \\
\hline
1 & 0 & 0 & 0 & 0 & 7/10 & 57/10 & -53/10 & 337/10
\end{array}\right]$$

Phase II applies

$$\left[\begin{array}{cccccccc|c}
0 & -1/8 & 0 & 0 & 1 & 1/8 & (1/4) & 0 & 4 \\
0 & 11/40 & 0 & 1 & 0 & -7/80 & 29/100 & 0 & 18/5 \\
0 & 5/4 & 0 & 0 & 0 & -1/4 & -3/2 & 1 & 1 \\
0 & 7/16 & 1 & 0 & 0 & 1/16 & -3/8 & 0 & 3/2 \\
\hline
1 & 53/8 & 0 & 0 & 0 & -5/8 & -9/4 & 0 & 39
\end{array}\right]$$

Phase II applies

$$\left[\begin{array}{cccccccc|c}
0 & -1/2 & 0 & 0 & 4 & 1/2 & 1 & 0 & 16 \\
0 & 191/200 & 0 & 1 & -27/25 & -191/800 & 0 & 0 & 268/25 \\
0 & 1/2 & 0 & 0 & 6 & 1/2 & 0 & 1 & 25 \\
0 & 1/4 & 1 & 0 & 3/2 & 1/4 & 0 & 0 & 15/2 \\
\hline
1 & 11/2 & 0 & 0 & 9 & 1/2 & 0 & 0 & 75
\end{array}\right]$$

The maximum value of P is 75 and it is achieved when
$x_1 = 0,\ x_2 = 15/2,\ x_3 = 268/25,\ s_1 = 0,\ s_2 = 16,\ s_3 = 25$

27. (Section 4.4)

We change our problem from minimizing $C = 5x_1 + 7x_2 + 6x_3$
to maximizing $z = -C = -5x_1 - 8x_2 - 6x_3$.

We write the constraints as

$$\begin{aligned}
x_1 + x_2 + x_3 &\le 10 \\
x_1 + x_2 + x_3 &\ge 10 \\
2x_1 + x_2 + 3x_3 &\le 19 \\
2x_2 + 3x_3 &\ge 21
\end{aligned}$$

Rewrite the constraints with $\le$

$$\begin{aligned}
x_1 + x_2 + x_3 &\le 10 \\
-x_1 - x_2 - x_3 &\le -10 \\
2x_1 + x_2 + 3x_3 &\le 19 \\
-2x_2 - 3x_3 &\le -21
\end{aligned}$$

Then

$$\begin{aligned}
x_1 + x_2 + x_3 + s_1 &= 10 \\
-x_1 - x_2 - x_3 + s_2 &= -10 \\
2x_1 + x_2 + 3x_3 + s_3 &= 19 \\
-2x_2 - 3x_3 + s_4 &= -21
\end{aligned}$$

$s_1 \ge 0,\ s_2 \ge 0,\ s_3 \ge 0,\ s_4 \ge 0$

27. (cont.)
The initial simplex tableaux is

x_1	x_2	x_3	s_1	s_2	s_3	s_4	
1	1	1	1	0	0	0	10
–1	–1	–1	0	1	0	0	–10
2	1	3	0	0	1	0	19
0	–2	(–3)	0	0	0	1	–21
5	7	6	0	0	0	0	0

Phase I applies

1	1/3	0	1	0	0	1/3	3
(–1)	–1/3	0	0	1	0	–1/3	–3
2	–1	0	0	0	1	1	–2
0	2/3	1	0	0	0	–1/3	7
5	3	0	0	0	0	2	–42

Phase I applies

0	0	0	1	1	0	0	0
1	1/3	0	0	–1	0	1/3	3
0	(–5/3)	0	0	2	1	1/3	–8
0	2/3	1	0	0	0	–1/3	7
0	4/3	0	0	5	0	1/3	–57

Phase I applies

0	0	0	1	1	0	0	0
1	0	0	0	–3/5	1/5	3/5	7/5
0	1	0	0	–6/5	–3/5	–1/5	24/5
0	0	1	0	4/5	2/5	–1/5	19/5
0	0	0	0	33/5	4/5	9/15	–317/5

The minimum value of C is 317/5 and it is achieved when
$x_1 = 7/5$, $x_2 = 24/5$, $x_3 = 19/5$, $s_1 = 0$, $s_2 = 0$, $s_3 = 0$, $s_4 = 0$

28. (Section 4.1)
Maximize z = 500A + 600B + 1000C
Subject to
$A \geq 0, B \geq 0, C \geq 0$
$2A + 2B + 3C \leq 500$
$4A + 5B + 5C \leq 1000$

P	A	B	C	s_1	s_2	
0	2	2	3	1	0	500
0	4	5	5	0	1	1000
0	–500	–600	–1000	0	0	0

28. (cont.)

$$\left[\begin{array}{cccccc|c} -0.4 & -1 & 0 & 1 & 0 & -0.6 & 200 \\ 0.8 & 1 & 1 & 0 & 0 & 0.20 & 200 \\ \hline 300 & 400 & 0 & 0 & 1 & 200 & 2000 \end{array}\right]$$

$P = 2,000$; $x_1 = 0$, $x_2 = 9$, and $x_3 = 200$

29. (Section 4.2)
Minimize $C = 5x_1 + 2x_2 + x_3$
Subject to
$x_1 \geq 0, x_2 \geq 0, x_3 \geq 0$
$3x_1 + 2x_2 + 5x_3 \geq 20$
$20x_1 + 10x_2 + 10x_3 \geq 40$

The dual is
Maximize $z = 20y_1 + 20y_2 + 40y_3$
Subject to
$y_1 \geq 0, y_2 \geq 0, y_3 \geq 0$
$3y_1 + 2y_2 + 20y_3 \leq 5$
$2y_1 + 3y_2 + 10y_3 \leq 2$
$5y_1 + 3y_2 + 10y_3 \leq 1$

CHAPTER 5

FINANCE

1. A resident of the United States has a base income of $23,000 after adjustments for deductions. New legislation by Congress would tax this income at a rate of 11 percent. What tax would be due?

2. Write .92 as a percent.

3. Write 32.4% as a decimal.

4. What percent of 45 is 5?

5. Sixteen percent of what number is 8?

In problems 6 – 11, find the simple interest. Assume all rates are on per annum basis.

6. $8,537 at 10% for 8 months.

7. $18,320 at 12% for 9 months.

8. $131,750 at 14.63% for 4 months.

9. $7,285 at 9.4% for 132 days.

10. $52,486 at 12.76% from July 8 of one year to October 15 of the next year. (365-day year)

11. $24,281 at 12 3/4% from June 8 of one year to February 14 of the next year.

In problems 12 – 14, find the proceeds for the given amount repaid, time, and simple discount rate.

12. $42,000, discount rate 12%, 6 months.

13. $16,200, discount rate 18%, 7 months.

14. $16,118, discount rate 14 1/2%, 95 days.

In problems 15 – 16, John borrowed $1,000 discounted at 10% for six months.

15. How much did he receive when the loan was made?

16. What annual rate of interest is he paying for the money actually received?

17. What is the ending balance from an initial deposit of $1,500 at 8% compounded annually for 4 years?

18. What is the ending balance from an initial deposit of $2,750 at 10% compounded semiannually for 4 years?

19. What is the ending balance from an initial deposit of $4,250 at 12% compounded quarterly for 6 years?

20. What is the ending balance from an initial deposit of $675.20 at 10 1/2% compounded semiannually for 5 years?

21. What is the ending balance from an initial deposit of $5,000 at 14% compounded monthly for 4 years?

In problems 22 – 26, find the amount of interest earned by the deposit given.

22. $2,000 at 9% compounded annually for 10 years.

23. $2,800 at 12% compounded semiannually for 8 years.

24. $1,262.25 at 14% compounded quarterly for 6 years.

25. $1,253.45 at 12% compounded quarterly for 26 quarters.

26. $1,609.25 at 13% compounded monthly for 90 months.

In problems 27 – 31, find the present value of the following amounts.

27. $5,000 in 5 years at 10% compounded annually.

28. $18,550 in 9 years at 8% compounded semiannually.

29. $12,400 in 8 years at 14% compounded quarterly.

30. $50,000 in 5 years and 6 months at 12% compounded quarterly.

31. $10,425 in 39 months at 13.75% compounded monthly.

In problems 32 – 33, suppose a job could be obtained with a $20,000 starting salary and a 5% yearly increase each year thereafter.

32. Give an expression for the salary during the seventh year.

33. Give an expression for the total gross income for the first 10 years.

34. If a five year $5,000 savings certificate earning 6% with quarterly compounding is bought, what is the amount of the investment at the end of five years provided no interest is withdrawn?

In problems 35 – 39, find the value of each of the following annuities.

35. $1,100 is deposited at the end of each year for 5 years; the interest rate is 11.5% compounded annually.

36. $275 is deposited at the end of each quarter for 3 years; the interest rate is 9.5% compounded quarterly.

37. $105 is deposited at the end of each month for 6 years; the interest is 10.5% compounded monthly.

38. $1,800 is deposited at the end of each six months for 5 years; the interest is 15% compounded semiannually.

39. $1,400 is deposited at the beginning of each year for 7 years; the interest is 12.5% compounded annually.

40. At the end of each three months Laura puts $200 into an account which pays 10% compounded quarterly. After 10 years she discontinues the payments but leaves the total amount in the account to collect interest for 2 more years. Determine the balance in the account at the end of 12 years.

In problems 41 – 44, determine the amount of each payment to be made to a sinking fund in order that enough money will be available to pay off the indicated loan.

41. $5,000 loan, money earns 14% compounded annually, 10 annual payments.

42. $120,000 loan, money earns 10% compounded semiannually, 8 1/2 years.

43. $49,200 loan, money earns 12.42% compounded quarterly, 12 1/4 years.

44. $29,000 loan, money earns 12% compounded annually, 42 months.

In problems 45 – 48, find the present value of each of the following ordinary annuities.

45. Annual payments of $1,000, 6 years, 10% compounded annually.

46. Monthly payments of $274.14, 48 months, 12% compounded monthly.

47. Semiannual payments of $1,471.06, 5 years, 13.5% compounded every six months.

48. Quarterly payments of $15,000, 6 1/2 years, 15.4% compounded quarterly.

49. Suppose a person borrows $15,000 to buy a car when the interest rate is one percent per month. How much should the equal monthly payments be if the loan is to be paid off in 60 months?

50. Tami's parents establish an ordinary annuity to accumulate $35,000 in fifteen years for her college expenses. The annuity earns 10% compounded quarterly. How much should their quarterly payments be?

51. An alumnus wants to set up a trust (which earns 9% interest compounded semiannually) to provide a grant to his alma mater of $10,000 every six months, beginning six months from now, for six years. How much should be deposited in the trust?

In problems 52 – 54, a $152,400 loan is taken out, at 11.5% for 25 years, for the purchase of a house. The loan requires monthly payments.

52. Find the amount of each payment.

53. Determine the total amount repaid over the life of the loan.

54. Find the total interest paid over the life of the loan.

55. A corporation may obtain a machine by leasing it for 6 years (the useful life) at an annual rent of \$3,300 or by purchasing the machine for \$12,000. Which alternative is preferable if the corporation can invest money at 8% per annum?

56. A bond has a face amount of \$1,000 and matures in 8 years. The nominal interest rate is 8.5%. What is the price of the bond to yield a true interest rate of 8%?

SOLUTIONS

1. (Section 5.1)
   (.11)(23,000) = \$2,530.00

2. (Section 5.1)
   92%

3. (Section 5.1)
   .324

4. (Section 5.1)
   $(x\%)(45) = 5$
   $x\% = 5/45$
   $x\% = .11\overline{1}$
   $x\% \approx 11.11$ or $11\ 1/9\%$

5. (Section 5.1)
   $(.16)(x) = 8$
   $x = 8/.16 = 50$

6. (Section 5.1)

   $$I = Prt = \$8{,}537(.10)\left(\frac{8}{12}\right) = \$569.13$$

7. (Section 5.1)

   $$I = \$18{,}320(.12)\left(\frac{9}{12}\right) = \$1{,}648.80$$

8. (Section 5.1)

   $$I = \$131{,}750(.1463)\left(\frac{4}{12}\right) = \$6{,}425.01$$

9. (Section 5.1)

   $$I = \$7{,}285(.094)\left(\frac{132}{365}\right) = \$247.65$$

10. (Section 5.1)

$$\$52,486(.1276)\left(\frac{464}{365}\right) = \$8,513.72$$

11. (Section 5.1)

$$I = \$24,281(.1275)\left(\frac{251}{365}\right) = \$2,128.91$$

12. (Section 5.1)

$$P = A(1 - dt) = \$42,000(1 - .12(.5)) = \$39,480$$

13. (Section 5.1)

$$P = \$16,200\left[1 - .18\left(\frac{7}{12}\right)\right] = \$14,499$$

14. (Section 5.1)

$$P = \$16,118\left[1 - .145\left(\frac{95}{365}\right)\right] = \$15,509.71$$

15. (Section 5.1)

$$P = \$1,000[1 - .10(.5)] = \$950$$

16. (Section 5.1)

$$r = \frac{I}{Pt} = \frac{50}{1000(.5)} = .10 = 10\% \text{ annual simple interest}$$

17. (Section 5.2)

$$A = \$1500(1 + .08)^4 = \$2040.73$$

18. (Section 5.2)

$$A = \$2750\left(1 + \frac{.10}{2}\right)^8 = \$4063$$

19. (Section 5.2)

$$A = \$4250\left(1 + \frac{.12}{4}\right)^{24} = \$8639.37$$

20. (Section 5.2)

$$A = \$675.20\left(1 + \frac{.105}{2}\right)^{10} = \$1126.38$$

21. (Section 5.2)

$$A = \$5000\left(1 + \frac{.14}{12}\right)^{48} = \$8725.03$$

22. (Section 5.2)

$$I = P[(1 + i)^{n} - 1] = \$2000[(1 + .09)^{10} - 1] = \$2734.73$$

23. (Section 5.2)

$$I = \$2800\left[\left(1 + \frac{12}{2}\right)^{16} - 1\right] = \$4312.98$$

24. (Section 5.2)

$$I = \$1262.25\left[\left(1 + \frac{.14}{4}\right)^{24} - 1\right] = \$1619.88$$

25. (Section 5.2)

$$I = \$1253.45\left[\left(1 + \frac{.12}{4}\right)^{26} - 1\right] = \$1449.73$$

26. (Section 5.2)

$$I = \$1609.25\left[\left(1 + \frac{.13}{12}\right)^{90} - 1\right] = \$2634.83$$

27. (Section 5.2)

$$P = A(1 + i)^{-n} = \$5000(1 + .10)^{-5} = \$3104.61$$

28. (Section 5.2)

$$P = \$18{,}550\left(1 + \frac{.08}{2}\right)^{-18} = \$9156.80$$

29. (Section 5.2)

$$P = \$12{,}400\left(1 + \frac{.14}{4}\right)^{-32} = \$4124.11$$

30. (Section 5.2)

$$P = \$50{,}000\left(1 + \frac{.12}{4}\right)^{-22} = \$26{,}094.63$$

31. (Section 5.2)

$$P = \$10{,}425\left(1 + \frac{.1375}{12}\right)^{-39} = \$6685.04$$

32. (Section 5.2)

$S_7 = \$20{,}000(1 + .05)^6 = \$26{,}801.91$

33. (Section 5.2)

Gross income equals the sum of $20{,}000(1 + .05)^n$ for $n = 0,1,\ldots,9$.

34. (Section 5.2)

$$A = \$5{,}000\left(1 + \frac{.06}{4}\right)^{20} = \$6734.28$$

35. (Section 5.3)

$$A = P \bullet A(ni) = P \bullet \frac{(1 + i)^n - 1}{i} = \$1{,}100 \bullet \frac{(1 + .115)^5 - 1}{.115} = \$6919.03$$

36. (Section 5.3)

$$A = \$275 \bullet \frac{\left(1 + \frac{.095}{4}\right)^{12} - 1}{\frac{.095}{4}} = \$3767.08$$

37. (Section 5.3)

$$A = \$105 \bullet \frac{\left(1 + \frac{.105}{12}\right)^{72} - 1}{\frac{.105}{12}} = \$10{,}469.67$$

38. (Section 5.3)

$$A = \$1800 \bullet \frac{\left(1 + \frac{.15}{2}\right)^{10} - 1}{\frac{.15}{2}} = \$25{,}464.76$$

39. (Section 5.3)

$$A = \$1400 \bullet \frac{(1 + .125)^8 - 1}{.125} = \$17{,}536.79$$

40. (Section 5.3)

$$A_{10} = \$200 \bullet \frac{\left(1 + \frac{.10}{4}\right)^{40} - 1}{\frac{.10}{4}} = \$13{,}480.51$$

$$A_{12} = \$13{,}480.51 \bullet \left(1 + \frac{.10}{4}\right)^8 = \$16{,}424.69$$

41. (Section 5.3)

$$P = \frac{A}{A(n,i)} = \frac{A \bullet i}{(1 + i)^n - 1} = \frac{5000(.14)}{(1 + .14)^{10} - 1} = \$258.57$$

42. (Section 5.3)

$$P = \frac{120{,}000\left(\frac{.10}{2}\right)}{\left(1 + \frac{.10}{2}\right)^{17} - 1} = \$4643.90$$

43. (Section 5.3)

$$P = \frac{49{,}200\left(\frac{.1242}{4}\right)}{\left(1 + \frac{.1242}{4}\right)^{49} - 1} = \$439.73$$

44. (Section 5.3)

$$P = \frac{29{,}000(.12)}{(1 + .12)^{3.5} - 1} = \$7148.20$$

45. (Section 5.4)

$$V = P \bullet P(n,i) = P \bullet \frac{1 - (1 + i)^{-n}}{i} = \$1000 \bullet \frac{1 - (1 + .10)^{-6}}{.10} = \$4355.26$$

46. (Section 5.4)

$$V = \$274.14 \bullet \frac{1-\left(1+\frac{.12}{12}\right)^{-48}}{\frac{.12}{12}} = \$10{,}410.18$$

47. (Section 5.4)

$$V = \$1471.06 \bullet \frac{1-\left(1+\frac{.135}{2}\right)^{-10}}{\frac{.135}{2}} = \$10{,}452.57$$

48. (Section 5.4)

$$V = \$15{,}000 \bullet \frac{1-\left(1+\frac{.154}{4}\right)^{-26}}{\frac{.154}{4}} = \$243{,}708.30$$

49. (Section 5.4)

$$P = \frac{A}{A(n,i)} = 15{,}000\left[\frac{1-(1+.01)^{-26}}{.01}\right]^{-1} = \$333.67$$

50. (Section 5.4)

$$P = 35{,}000\left[\frac{1-(1+.10/4)^{-60}}{.10/4}\right]^{-1} = \$1{,}132.37$$

51. (Section 5.4)

$$V = P \bullet P(n,i) = P \bullet \frac{1-(1+i)^{-n}}{i}$$

$$= \$10{,}000 \bullet \frac{1-\left(1+\frac{.09}{2}\right)^{-12}}{\frac{.09}{2}} = \$91{,}185.81$$

52. (Section 5.4)

$$P = \frac{V}{P(n,i)} = \frac{V \bullet i}{1-(1+i)^{-n}} = \frac{\$152{,}400\left(\frac{.115}{12}\right)}{1-\left(1+\frac{.115}{12}\right)^{-300}} = \$1549.10$$

53. (Section 5.4)

    $T = 300(\$1549.10) = \$464{,}729.61$

54. (Section 5.4)

    $I = \$464{,}729.61 - 152{,}400 = \$312{,}329.61$

55. (Section 5.5)

    $P = 3000[(1.08)^{-6} + (1.08)^{-5} + (1.08)^{-4} + (1.08)^{-3} + (1.08)^{-2} + (1.08)^{-1}] > 12{,}000$
    The purchase is a better choice.

56. (Section 5.6)

    $(1000)(\frac{1}{2})(0.085) = 42.50$
    $(42.50)(16) = 680$

    $$42.50\,\frac{(1 - (.04)^{-16})}{.04} = 495.22$$

CHAPTER 6

SETS; COUNTING TECHNIQUES

In problems 1 – 8, let $U = \{1,2,3,4,5,6,7,8,9,10\}$, $A = \{2,6,10\}$, $B = \{1,3,5\}$, and $D = \{1,3,6,7,10\}$. Give the elements in the following problems.

1. $\overline{D} \cup B$

2. $\overline{A \cap D}$

3. $D \cap (A \cup B)$

4. $\overline{A \cup B}$

5. $\overline{\varnothing}$

6. $c\,(A \cup B)$

7. $c\,(\overline{A})$

8. $c\,(\overline{A \cup B})$

In problems 9-15, use Venn Diagrams to indicate which portion represents the following set.

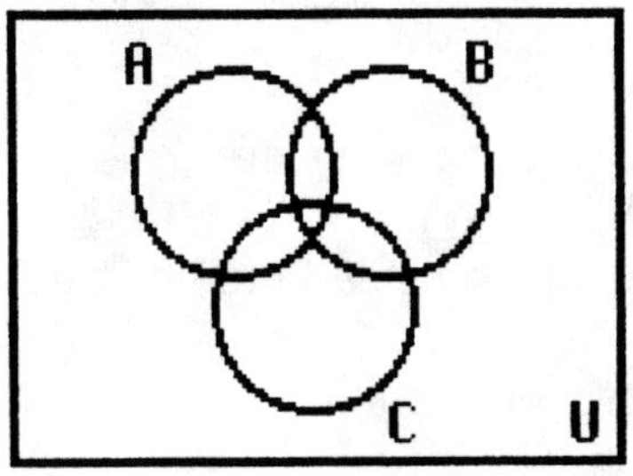

9. $\overline{A} \cup \overline{B}$

10. $\overline{A} \cap B$

11. $\overline{A \cup B}$

12. $A \cap (B \cup C)$

13. $\overline{C} \cap (A \cup B)$

14. $\overline{(A \cup C)} \cap B$

15. $(A \cap B) \cup (A \cap C)$

In problems 16 – 21,

Let $U = \{2,4,6,8, \ldots\}$ $\quad A = \{2,6,10\}$ $\quad B = \{4,6\}$

$C = \{2,6,8,10,12, \ldots\}$ $\quad D = \{4,8,12,16,20\}$ $\quad E = \varnothing$

16. Find subsets of A.

17. Find proper subsets of B.

18 Find $\overline{C}$.

19. Find number of subsets of D.

20. Find c (E)

21. Find $\left[\,\overline{\overline{(A \cap B)} \cap E}\,\right]$

22. $c(P) = 3$, $c(N) = 4$ and $c(P \cap N) = 2$
Find $c(P \cup N)$

In problems 23 – 28, a survey of residents in a community showed the following information about their lunch eating habits:

29 ate salads
26 ate sandwiches
25 ate sweets
13 ate both a salad and a sandwich
11 ate both a sandwich and sweets
18 ate both a salad and sweets
6 ate all three types
7 ate none of them

23. How many of these students ate a sandwich but not a salad nor sweets?

24. How many ate sandwiches and sweets but not a salad?

25. How many ate only a salad?

26. How many ate nothing or only sweets?

27. How many ate sweets and salad but not a sandwich?

28. How many residents were surveyed?

For problems 29 – 33, a survey of freshman calculus students at a certain university showed the following graphics calculator use:

21 use TI	13 use TI and Casio
45 use Casio	14 use HP and TI
24 use HP	15 use Casio and HP
1 uses none of the above	8 use all three of the above

29. How many students were surveyed?

30. How many of the Casio users do NOT use TI?

31. How many students use at least one calculator besides Casio?

32. How many students use at most two of the calculators?

33. How many students use exactly one of the calculators?

34. In how many ways is it possible for five persons to line up at a check-out counter?

35. In a marketing survey consumers are asked to give their first three choices, in order of preference, of 9 different drinks. In how many different ways can they indicate their choices?

36. A class consists of 15 students. The instructor wants to pick a group of 4 to work on a special project.
(a) How many different groups of 4 can he choose?
(b) If the class consists of 10 girls and 5 boys, how many different groups of 4 are made up of 2 boys and 2 girls?

37. Twelve coins must be placed in three boxes A,B,C. In how many ways can this be done if 4 coins must be placed in A, 3 in B, and 5 in C?

38. On a single shelf we are to arrange 4 computer science books and 5 mathematics books. In how many ways can this be done if the computer science books are to be grouped together and the mathematics books are also to be together?

For problems 39 – 41, expand by the binomial theorem.

39. $(2w + 3u)^5$

40. $(2x - 3y)^4$

41. $(100x - 1)^5$

42. Find the fifth term in the expression $(b - 3a)^{15}$.

43. What is the coefficient of x^7 in the expansion of $(x + y)^{12}$?

44. How many subsets can be chosen from a set with 4 elements?

45. Show that $$\binom{n+1}{k} = \binom{n}{k-1} + \binom{n}{k}$$

using only the fact that $$\binom{n}{k} = \frac{n!}{k!\,(n-k)!}$$

46. How many different collections can be formed from 8 different books, if each collection has 3 or more books in it?

47. In how many distinct arrangements can 8 children form a circle to play ring-around-the-rosy?

48. In how many ways can 8 differently colored interlocking beads be put together to form a (closed) necklace?

49. Simplify: $$\frac{92!\ 3!}{(100-10)!\ 7!}$$

50. P(8, 3) =

51. C(5, 2) =

For problems 52 – 55, an experiment consists of tossing a fair coin 6 times.

52. How many different outcomes are possible?

53. How many different outcomes have exactly 3 heads?

54. How many outcomes have at most 2 tails?

55. How many outcomes have at least 3 tails?

56. How many different 12-letter words (real or imaginary) can be formed from the letters in the word APPLICATIONS?

57. There are six houses on a block, each with three outer doors. A person knocks on a door at random. How many possible doors can he select?

58. A person has 3 pairs of shoes and eight different pairs of socks. In how many ways can one of each be selected?

59. A menu has 5 soups, 3 salads, 6 main courses, 2 vegetables, and 3 desserts. In how many ways can one of each be chosen?

60. A coffee shop offers 6 types of coffee, 2 different creamers, and 8 pastries. In how many ways can one of each be selected?

61. A student has a 10-question multiple choice test with five answers for each question. In how many ways can the student "guess" at all of the answers?

62. How many license plates can a state have if each license plate has 2 letters followed by 4 digits, if the first digit cannot be zero?

63. How many 6-digit even numbers are there?

SOLUTIONS

1. (Section 6.1)
{1,2,3,4,5,8,9}

2. (Section 6.1)
{1,2,3,4,5,7,8,9}

3. (Section 6.1)
{1,3,6,10}

4. (Section 6.1)
{4,7,8,9}

5. (Section 6.1)
U = {1,2,3,4,5,6,7,8,9,10}

6. (Section 6.2)
6

7. (Section 6.2)
7

8. (Section 6.2)
4

9. (Section 6.1)

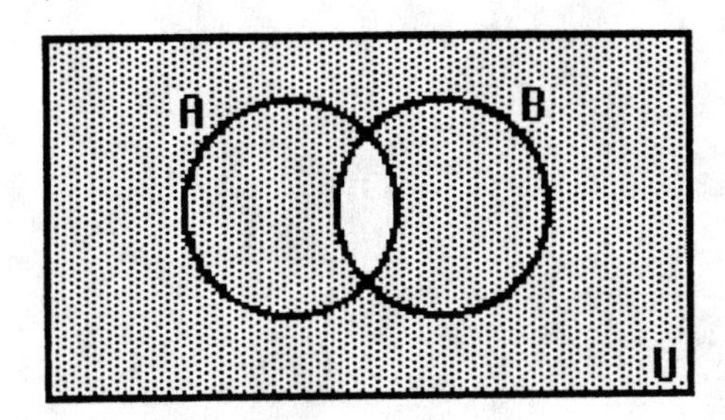

10. (Section 6.1)

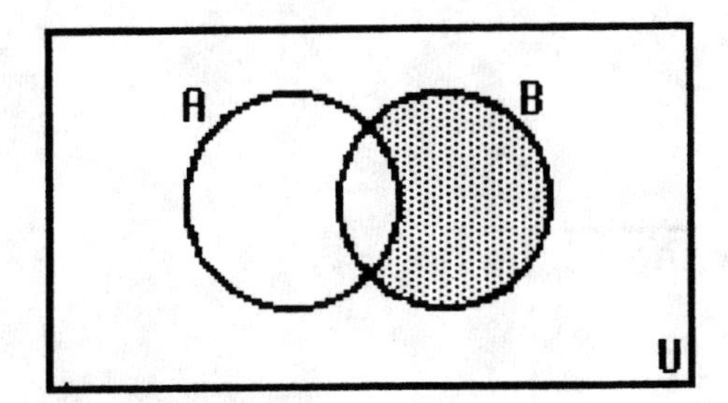

11. (Section 6.1)

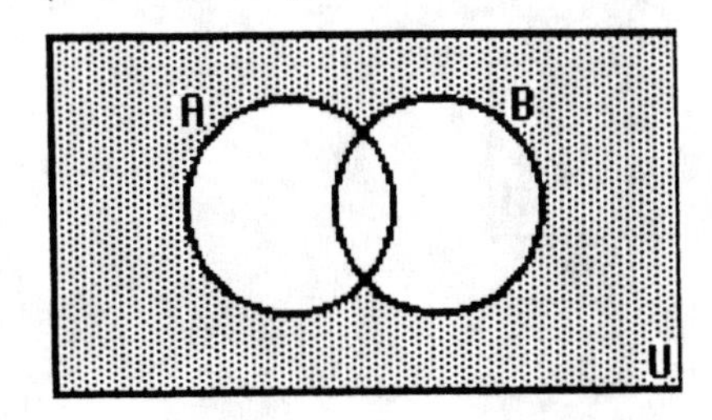

12. (Section 6.1)

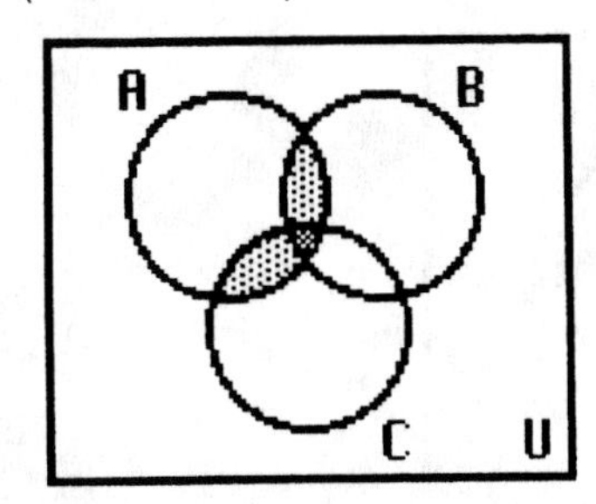

13. (Section 6.1)

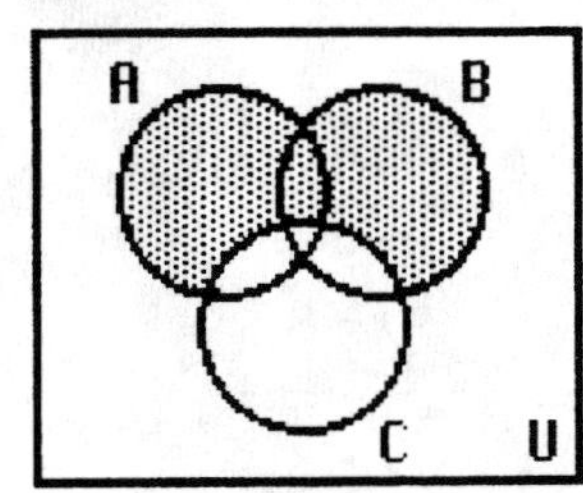

14. (Section 6.1)

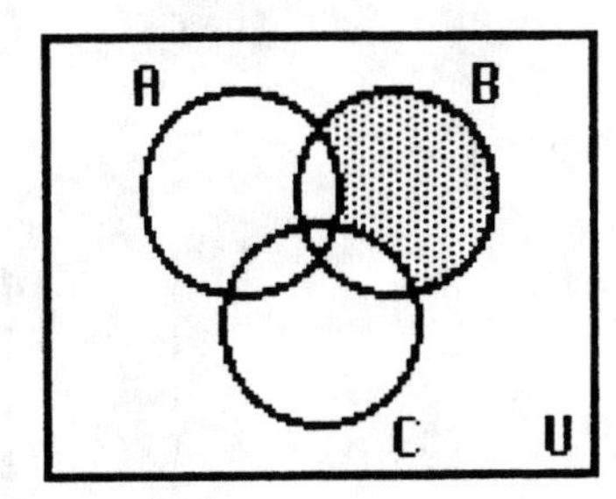

15. (Section 6.1)

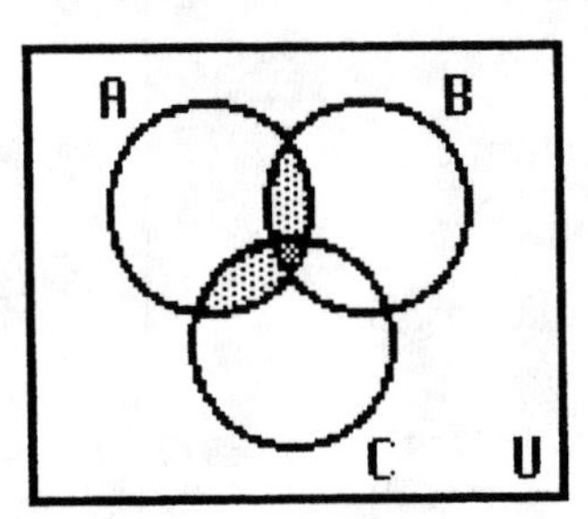

16. (Section 6.1)
$\emptyset$, {2}, {6}, {10}, {2,6}, {2,10}, {6,10}, {2,6,10}

17. (Section 6.1)
$\emptyset$, {4}, {6}

18. (Section 6.1)
{4}

19. (Section 6.6)
$2^5 = 32$

20. (Section 6.6)
    0

21. (Section 6.1)

    $(\overline{A \cap B} \cap E) = \emptyset$

    Thus $\overline{(\overline{A \cap B} \cap E)} = U = \{2,4,6,8, \ldots\}$

22. (Section 6.2)
    $c(P \cup N) = 5$ $\quad c(P) + c(N) - c(P \cap N) = 3 + 4 - 2 = 5$

23. (Section 6.2)
    8

24. (Section 6.2)
    5

25. (Section 6.2)
    4

26. (Section 6.2)
    9

27. (Section 6.2)
    12

28. (Section 6.2)
    51

29. (Section 6.2)
    57

30. (Section 6.2)
    32

31. (Section 6.2)
    11

32. (Section 6.2)
    49

33. (Section 6.2)
    30

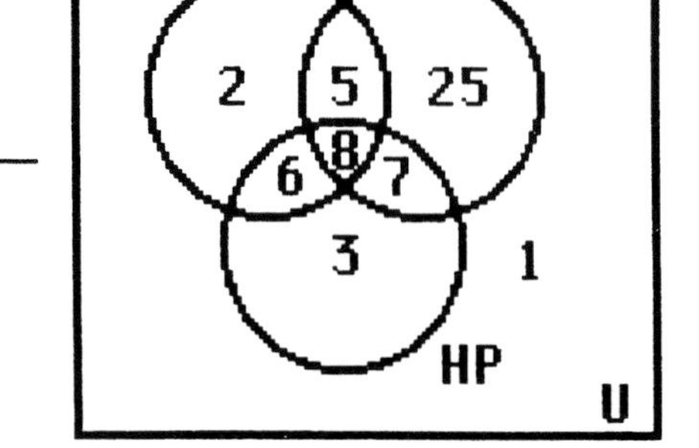

34. (Section 6.4)
    $5! = 5 \cdot 4 \cdot 3 \cdot 2 \cdot 1 = 120$

35. (Section 6.6)
    $\binom{9}{1} \cdot \binom{8}{1} \cdot \binom{7}{1} = 9 \cdot 8 \cdot 7 = 504$

36. (Section 6.6)

(a) $\binom{15}{4} = \dfrac{15!}{4!\,11!} = 1365$

(b) $\binom{5}{2} \cdot \binom{10}{2} = \dfrac{5!}{2!\,3!} \bullet \dfrac{10!}{2!\,8!} = 450$

37. (Section 6.6)

$$\binom{12}{4} \cdot \binom{8}{3} \cdot \binom{5}{5} = 27{,}720$$

38. (Section 6.6)
$4! \bullet 5! \bullet 2 = 5760$

39. (Section 6.7)

$$(2w + 3u)^5 = \binom{5}{0}(2w)^5 + \binom{5}{1}(2w)^4(3u) + \binom{5}{2}(2w)^3(3u)^2$$

$$+ \binom{5}{3}(2w)^2(3u)^3 + \binom{5}{4}(2w)(3u)^4 + \binom{5}{5}(3u)^5$$

$$= 32w^5 + 240w^4u + 720w^3u^2 + 1080w^2u^3 + 810wu^4 + 243u^5$$

40. (Section 6.7)
$(2x - 3y)^4 = 16x^4 - 96x^3y + 216x^2y^2 - 216xy^3 + 81y^4$

41. (Section 6.7)
$(100x - 1)^5 = 100^5x^5 - 5(100)^4x^4 + 10(100)^3x^3 - 10(100)^2x^2 + 500x - 1$

42. (Section 6.7)

$$\binom{15}{4} b^{11}(-3a)^4 = 110{,}565b^{11}a^4$$

43. (Section 6.7)

$$\binom{12}{5} = 792$$

44. (Section 6.6)
$2^4 = 16$

45. (Section 6.7)

$$\binom{n}{k-1} + \binom{n}{k} = \frac{n!}{(k-1)!\,(n-k+1)!} + \frac{n!}{k!\,(n-k!)} =$$

$$\frac{n!k}{k!\,(n-k+1)!} + \frac{n!\,(n-k+1)}{k!\,(n-k+1)!} = \frac{n!\,[k+n-k+1]}{k!\,(n-k+1)!} = \frac{(n+1)!}{k!\,(n+1-k)!} = \binom{n+1}{k}$$

46. (Section 6.6)

$$\binom{8}{3} + \binom{8}{4} + \binom{8}{5} + \binom{8}{6} + \binom{8}{7} + \binom{8}{8} = 219$$

47. (Section 6.6)

$$\binom{8}{1} \cdot \binom{7}{1} \cdot \binom{6}{1} \cdot \binom{5}{1} \cdot \binom{4}{1} \cdot \binom{3}{1} \cdot \binom{2}{1} \cdot \binom{1}{1} = 40{,}320$$

48. (Section 6.6)
$8! = 40{,}320$; $8! \div 8 = 5{,}040$

49. (Section 6.4)

$$\frac{92!\ 3!}{(100-10)!\ 7!} = \frac{92!\ 3!}{90!\ 7!} = \frac{92 \bullet 91}{7 \bullet 6 \bullet 5 \bullet 4} = \frac{299}{30}$$

50. (Section 6.4)
$P(8,3) = 8 \bullet 7 \bullet 6 = 336$

51. (Section 6.5)

$$C(5,2) = \frac{5!}{(5-2)!\ 2!} = \frac{5 \bullet 4}{2 \bullet 1} = 10$$

52. (Section 6.6)

$2^6 = 64$

53. (Section 6.5)
$C(6,3) = 20$

54. (Section 6.6)
$C(6,0) + C(6,1) + C(6,2) = 22$

55. (Section 6.6)
$64 - 22 = 42$

56. (Section 6.6)

$$\frac{12!}{2!\ 2!\ 2!} = 59{,}875{,}200$$

57. (Section 6.3)
$(6)(3) = 18$

58. (Section 6.3)
$(3)(8) = 24$

59. (Section 6.3)
$(5)(3)(6)(2)(3) = 540$

60. (Section 6.3)
$(6)(2)(8) = 96$

61. (Section 6.3)
$5^{10} = 9{,}765{,}625$

62. (Section 6.3)
$(26)^2(9)(10)^3 = 6{,}084{,}000$

63. (Section 6.3)
$9(10)^4(5) = 450{,}000$

CHAPTER 7

PROBABILITY

In problems 1 – 4, write a sample space for each experiment.

1. A die is rolled.

2. A coin is tossed 2 times.

3. A student is given a 3-question True-False test.

4. A die is rolled and at the same time a coin is tossed.

In problems 5 – 9, let $S = \{e_1, e_2, e_3, e_4, e_5\}$, $E = \{e_1, e_3, e_5\}$, $F = \{e_1, e_2, e_4\}$, $P(e_1) = 0.1$, $P(e_2) = 0.2$, $P(e_3) = 0.35$, $P(e_4) = 0.15$, $P(e_5) = 0.2$

Determine:

5. $P(E)$

6. $P(F)$

7. $P(E \cup F)$

8. $P(\overline{E \cap F})$

9. Are E and F mutually exclusive? Justify.

In problems 10 – 13, a box contains 5 marbles numbered 1, 2, 3, 4, and 5. An experiment consists of a fair coin once and then choosing a marble at random from the box. Write a set for:

10. The sample space.

11. Event E, the marble is numbered 2 or 5.

12. Event F, the coin shows heads.

13. Are the outcomes in the sample space equally likely?

In problems 14 – 18, If $P(E) = 0.4$, $P(F) = 0.5$, and $P(E \cap F) = 0.2$

14. Determine $P(\bar{E})$

15. Determine $P(E \cup F)$

16. Determine $P(E \mid F)$

17. Are E and F mutually exclusive? Explain.

18. Are E and F independent? Explain.

In problems 19 – 27, a single card is drawn at random from a deck of 52 cards. Jack, Queen, and King are considered the only picture cards. Find the probability that the card will be:

19. A heart.

20. A diamond 7.

21. A picture card.

22. Black or picture card.

23. The cards from 3 to 7 inclusive.

24. Black card, given it is a picture card.

25. A king, given it is a picture card.

26. A picture card, given it is a queen.

27. A picture card, given it is a 7.

28. In the game Over and Under, a pair of dice is rolled and one can bet whether the sum of dots showing on the dice is over 7, under 7, or exactly 7. What are the odds in favor of you winning when you bet that over 7 wins?

29. If a card is drawn at random from an ordinary deck of playing cards, what are the odds in favor of it being a picture card (king, queen, jack)?

30. The odds in favor of an event are 3 to 7. What is the probability that the event will occur?

In problems 31 – 32, a store owner has observed that the number of customers coming to the store during 100 store days was distributed as follows:

number of customers	0	1	2	3	4	5 or more
number of days	5	36	10	15	22	12

On this basis, what is the probability of the following customer distributions?

31. At least one

32. At most 2

In problems 33 – 42, find the probability for the following sums in a throw of two fair dice.

33. 3

34. 7

35. 11

36. At least 7

37. No more than 4

38. Odd and less than 3

39. 4, given that the sum is less than 6

40. 8, given that the one die is 5

41. 8, given that the one die is odd

42. At least 8, given that one die is at least 4.

In problems 43 – 46, an urn contains 3 white balls and 5 red balls. Two balls are drawn at random. If the first is replaced before the second is drawn, what is the probability that:

43. Two red balls are drawn?

44. The first ball is red and the second is white?

45. The first ball is white and the second is red?

46. One of the balls is white and one is red?

For problems 47 – 53, two cards are drawn at random from an ordinary deck of playing cards. The first is not replaced before the second is drawn. What is the probability that:

47. Both cards are aces?

48. Both cards are red?

49. At least one card is black?

50. At least one card is a heart?

51. The second card is black, given that the first card is red?

52. The second card is 7, given the first card was the 2 of diamonds?

53. The second card is a picture, given that the first card was not a picture?

54. A jar contains 8 red marbles, 9 blue marbles, and 6 green marbles. Two marbles are picked at random. What is the probability that both are blue?

55. A jar contains 8 red marbles, 9 blue marbles, and 6 green marbles. Two marbles are picked at random. What is the probability that at least one is blue?

56. A jar contains 8 red marbles, 9 blue marbles, and 6 green marbles. Two marbles are picked at random. What is the probability that one is green and one is blue?

57. A fair coin is tossed 8 times. Find the probability it comes up heads exactly 6 times.

58. A fair coin is tossed 8 times. Find the probability it comes up heads at least twice.

59. Five cards are dealt at random from a deck of 52 playing cards. Find the probability that three are sevens.

SOLUTIONS

1. (Section 7.1)
   {1,2,3,4,5,6}

2. (Section 7.1)
   {HH,HT,TH,TT}

3. (Section 7.1)
   {TTT,TFF,FTF,FFT,TTF,TFT,FTT,FFF}

4. (Section 7.1)
   {1H,1T,2H,2T,3H,3T,4H,4T,5H,5T,6H,6T}

5. (Section 7.1)
   $P(E) = .5$

6. (Section 7.1)
   $P(F) = .45$

7. (Section 7.1)
   $P(E \cup F) = 1.00$

8. (Section 7.1)
   $P(\overline{E \cap F}) = .9$

9. (Section 7.2)
   No; both sets contain event e_1.

10. (Section 7.1)
    S = {H1,T1,H2,T2,H3,T3,H4,T4,H5,T5}

11. (Section 7.1)
    E = {H2,T2,H5,T5}

12. (Section 7.1)
    F = {H1,H2,H3,H4,H5}

13. (Section 7.2)
    Yes

14. (Section 7.2)
    $P(\bar{E}) = .6$

15. (Section 7.2)
$P(E \cup F) = .4 + .5 - .2 = .7$

16. (Section 7.4)
$P(E \mid F) = .2/.5 = .4$

17. (Section 7.2)
No; $(E \cap F) \neq \emptyset$

18. (Section 7.5)
Yes; $P(E \mid F) = P(E)$

19. (Section 7.2)
1/4

20. (Section 7.2)
1/52

21. (Section 7.2)
$\frac{12}{52} = \frac{3}{13}$

22. (Section 7.2)
$\frac{32}{52} = \frac{8}{13}$

23. (Section 7.2)
$\frac{20}{52} = \frac{5}{13}$

24. (Section 7.5)
1/2

25. (Section 7.5)
1/3

26. (Section 7.5)
1

27. (Section 7.5)
0

28. (Section 7.3)
$P = \frac{15}{36} = \frac{5}{12}$; odds 5 to 7

29. (Section 7.2)
$P = \frac{12}{52} = \frac{3}{13}$; odds of 3 to 10

30. (Section 7.1)
$P = \frac{3}{10}$

31. (Section 7.1)
$\frac{95}{100} = .95$

32. (Section 7.1)
$\frac{51}{100} = .51$

33. (Section 7.2)
$\frac{2}{36} = \frac{1}{18}$

34. (Section 7.2)
$\frac{6}{36} = \frac{1}{6}$

35. (Section 7.2)
$\frac{2}{36} = \frac{1}{18}$

36. (Section 7.2)
$\frac{21}{36} = \frac{7}{12}$

37. (Section 7.2)
$\frac{6}{36} = \frac{1}{6}$

38. (Section 7.2)
0

39. (Section 7.4)
$\frac{3}{10}$

40. (Section 7.4)
$\frac{2}{11}$

41. (Section 7.4)
$\frac{2}{29}$

42. (Section 7.4)
$\frac{5}{11}$

43. (Section 7.5)
$$\frac{5}{8} \bullet \frac{5}{8} = \frac{25}{64}$$

44. (Section 7.5)
$$\frac{5}{8} \bullet \frac{3}{8} = \frac{15}{64}$$

45. (Section 7.5)
$$\frac{3}{8} \bullet \frac{5}{8} = \frac{15}{64}$$

46. (Section 7.5)
$$\frac{5}{8} \bullet \frac{3}{8} + \frac{3}{8} \bullet \frac{5}{8} = \frac{30}{64} = \frac{15}{32}$$

47. (Section 7.5)
$$\frac{\binom{4}{2}\binom{48}{0}}{\binom{52}{2}} = \frac{1}{221}$$

48. (Section 7.5)
$$\frac{\binom{26}{2}\binom{26}{0}}{\binom{52}{2}} = \frac{25}{102}$$

49. (Section 7.5)
$$\frac{\binom{26}{1}\binom{26}{1}}{\binom{52}{2}} + \frac{\binom{26}{0}\binom{26}{2}}{\binom{52}{2}} = \frac{77}{102}$$

50. (Section 7.5)
$$\frac{\binom{13}{1}\binom{39}{1}}{\binom{52}{2}} + \frac{\binom{13}{2}\binom{39}{0}}{\binom{52}{2}} = \frac{15}{34}$$

51. (Section 7.5)
$$\frac{26}{51}$$

52. (Section 7.5)

$$\frac{4}{51}$$

53. (Section 7.5)

$$\frac{12}{51}$$

54. (Section 7.3)

$$\frac{C(9,2)}{C(23,2)} = \frac{36}{253}$$

55. (Section 7.3)

$$1 - \frac{C(14,2)}{C(23,2)} = \frac{162}{253}$$

56. (Section 7.3)

$$\frac{C(9,1)C(6,1)}{C(23,2)} = \frac{54}{253}$$

57. (Section 7.3)

$$\frac{C(8,6)}{2^8} = \frac{7}{64}$$

58. (Section 7.3)

$$1 - \left(\frac{1}{2^8} + \frac{8}{2^8}\right) = \frac{247}{256}$$

59. (Section 7.3)

$$\frac{C(4,3)C(48,2)}{C(52,5)} = \frac{4,512}{2,598,960} = 0.001736079$$

CHAPTER 8

ADDITIONAL PROBABILITY TOPICS

In problems 1 – 5, find the indicated probabilities by referring to the tree diagram shown below:

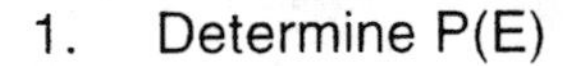

1. Determine P(E)

2. Determine $P(E \cup F)$

3. Determine $P(E \mid F)$

4. Are E and F mutually exclusive? Explain.

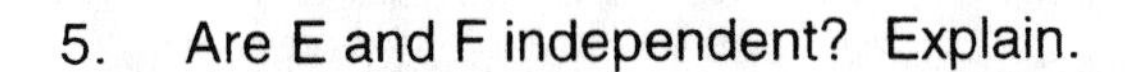

5. Are E and F independent? Explain.

6. An experimental test to detect a particular type of cancer indicates the presence of cancer in 90% of individuals known to have this type of cancer and in 15% of individuals known to be cancer-free. One hundred individuals volunteer to take the test. Of the 100, 60 are known to have the cancer and 40 are known to be cancer-free. If the test indicates that one of the individuals, randomly chosen, has the cancer, what is the probability that he or she is cancer-free?

7. Cystic fibrosis results from the presence of two recessive genes. If the parents are apparently normal, but each has a single gene for the disease, what is the probability that their first child will (a) have the disease, (b) be a carrier for the disease?

8. If a person buys ten tickets at $1 each in a lottery in which 1,000 tickets are sold and the prize is $500, what are his expected net winnings?

9. A lottery has a first prize of $10,000, a second prize of $1,000, and a third prize of $100. Two million tickets at $1 are sold. Find the expected winnings of a person buying 1 ticket.

For problems 10 – 11, determine the expected value of each of the following when a pair of fair dice are rolled.

10. The difference between two numbers showing, subtracting the smaller number from the larger.

11. The larger of the two numbers showing. (If both show the same number, that number is the larger.)

For problems 12 – 13, urn A contains 3 white and 5 black balls and urn B contains 2 white and 6 black balls. An experiment consists of tossing a biased coin (P(H) = 0.6) once and then choosing a ball at random from one of the urns; the ball is chosen from A if heads turns up and from B otherwise.

12. What is the probability that a white ball is drawn?

13. If the ball drawn is white, what is the probability that it came from urn A?

Use the following information for problems 14 – 17:

In the game Over-and-Under, a pair of dice is rolled and one bets $1 whether the sum of dots showing on the two dice is *over 7, under 7, or exactly 7* with payoffs of $2, $2, and $5 respectively. Determine your expected net winnings if you play as indicated.

14. Over 7

15. Under 7

16. Exactly 7

17. Is the game fair?

For problems 18 – 19, 3 cards are drawn (without replacement) from a standard deck of 52 cards.

18. What is the expected number of hearts?

19. If someone offers to pay you $100 for getting all 3 hearts, what is a fair price to pay for the game?

20. Each morning a manager has 6 dozen donuts delivered to his coffee shop. For each dozen he sells he makes a profit of $6.00; for each dozen that he does not sell he takes a loss of $3.00. He does not keep any overnight to sell the next day. Past experience indicates that the probability of selling the donuts is given by the following table.

Number of Dozens	Probability
6	0.05
5	0.15
4	0.20
3	0.30
2	0.10
1	0.10
0	0.10

Note: 0.10 is the probability that he will sell 1 dozen on any day, and so forth. Determine the expected number he sells and the expected number he does not sell.

21. A box contains 4 defective and 8 good bulbs. If 3 bulbs are selected at random, what is the probability that 1 is good?

For problems 22 – 23, a die is rolled 6 times. What is the probability of obtaining:

22. Exactly four 5's?

23. At least four 5's?

For problems 24 – 27, four percent of the items coming off an assembly line are defective. If the defective items occur randomly and ten items are chosen for inspection, what is the probability that:

24. Exactly two items are defective?

25. At most two items are defective?

26. No items are defective?

27. No more than three items are defective?

28. A man claims to be able to distinguish between scotch & bourbon 80% of the time. A test of 15 samples is given to him and, if he is correct at least 12 times, he proves his claim. What is the probability his claim is justified, but he does not pass?

29. A pair of dice are thrown once and their sum is the random variable X. What is the sample space and the associated probabilities?

30. Three fair coins are tossed and their "sum" is obtained by assigning values of Heads = 1 and Tails = 0. What is the sample space and the associated probabilities?

For problems 31 – 33, find the indicated probabilities by referring to the tree diagram shown below.

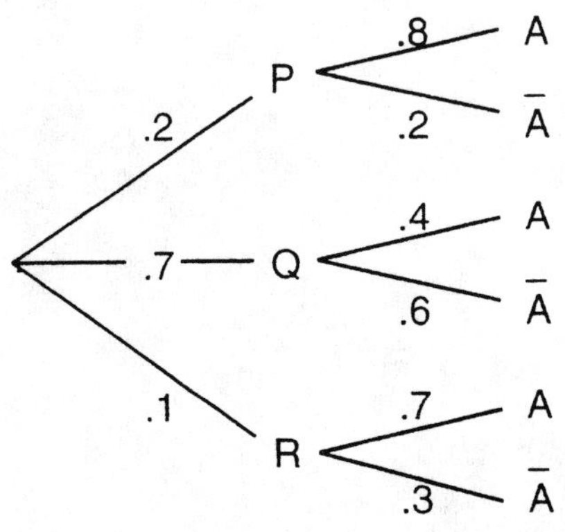

31. $P(A \mid P)$

32. $P(A)$

33. $P(\bar{A} \mid Q)$

34. Calculate the following binomial probability:
b(9, 6; 90) =

35. If A_1, A_2, A_3 form a partition of a sample space S and if $P(A_1) = .2$, $P(A_2) = .4$, $P(A_3) = .4$, and if E is an event of S for which $P(E \mid A_1) = .01$, $P(E \mid A_2) = .02$, $P(E \mid A_3) = .01$, find $P(A_1 \mid E)$.

36. From past history, a student has been determined to have a probability of studying on any given night of 0.4, a probability of going to the mall of 0.6, and a probability of doing both of 0.2. What is the probability that on a given night the student does neither?

37. A six-sided die with sides numbered 1 through 6 is rolled ten times. What is the probability of having 4 come up exactly 9 times?

38. A six-sided die with sides numbered 1 through 6 is rolled ten times. What is the probability of having 4 come up at least 9 times?

SOLUTIONS

1. (Section 8.1)
   $P(\bar{E}) = .6$

2. (Section 8.1)
   $P(E \cup F) = .4 + .5 - .2 = .7$

3. (Section 8.1)
   $P(E \mid F) = \frac{.2}{.5} = .4$

4. (Section 8.1)
   $P(E \cap F) \neq \phi$

5. (Section 8.1)
   Yes; $P(E \mid F) = P(E)$

6. (Section 8.1)
   $\frac{(.15)(.4)}{(.15)(.4) + (.9)(.6)} = .1$

7. (Section 8.1)
   (a) 1/4 (b) 1/2

8. (Section 8.3)
   $10\left(\frac{1}{1000}\ \$500\right) = \$5$

9. (Section 8.3)
   $\frac{1}{2{,}000{,}000} \bullet \$10{,}000 + \frac{1}{2{,}000{,}000} \bullet \$1{,}000 + \frac{1}{2{,}000{,}000} \bullet \$100 = \$.006 = .6¢$

10. (Section 8.3)
    $\frac{10}{36} + 2 \bullet \frac{8}{36} + 3 \bullet \frac{6}{36} + 4 \bullet \frac{4}{36} + 5 \bullet \frac{2}{36} = \frac{35}{18}$

11. (Section 8.3)

$$1 \bullet \frac{1}{36} + 2 \bullet \frac{3}{36} + 3 \bullet \frac{5}{36} + 4 \bullet \frac{7}{36} + 5 \bullet \frac{9}{36} + 6 \bullet \frac{11}{36} = \frac{161}{36}$$

12. (Section 8.1)

$$\frac{6}{10} \bullet \frac{3}{8} + \frac{4}{10} \bullet \frac{2}{8} = \frac{13}{40}$$

13. (Section 8.1)

$$\frac{\frac{3}{8} \bullet \frac{6}{10}}{\frac{3}{8} \bullet \frac{6}{10} + \frac{2}{8} \bullet \frac{4}{10}} = \frac{9}{13}$$

14. (Section 8.3)

P(Over 7) = $\frac{15}{36}$; Expected winnings = $\frac{15}{36} \bullet \$2 = \$.83$

15. (Section 8.3)

P(Under 7) = $\frac{15}{36}$; Expected winnings = $\frac{15}{36} \bullet \$2 = \$.83$

16. (Section 8.3)

P(Exactly 7) = $\frac{6}{36}$; Expected winnings = $\frac{6}{36} \bullet \$5 = \$.83$

17. (Section 8.3)

$E = \frac{15}{36} \bullet \$2 = \frac{15}{36} \bullet \$2 = \frac{6}{36}\ \$5 = \$.83$ for all bets, thus the game is not fair.

18. (Section 8.3)

$$0 \bullet \frac{\binom{13}{0}\binom{39}{3}}{\binom{52}{3}} + 1 \bullet \frac{\binom{13}{1}\binom{39}{2}}{\binom{52}{3}} + 2 \bullet \frac{\binom{13}{2}\binom{39}{1}}{\binom{52}{3}} + 3 \bullet \frac{\binom{13}{3}\binom{39}{0}}{\binom{52}{3}}$$

$$= \frac{3}{4}$$

19. (Section 8.3)

P(Winning) = $\frac{\binom{13}{3}\binom{39}{0}}{\binom{52}{3}} = \frac{11}{850}$; Cost = $\frac{11}{850} \bullet \$100 = \$129$

20. (Section 8.3)

Expected sales $= 0 \cdot (.10) + 1 \cdot (.10) + 2 \cdot (.10) + 3 \cdot (.30) + 4 \cdot (.20) + 5 \cdot (.15) + 6 \cdot (.05)$
$= 3.05$ dozen

Expected unsold $= 6 - 3.05 = 2.95$ dozen

21. (Section 8.1)

$$\frac{\binom{8}{1}\binom{4}{2}}{\binom{12}{3}} = \frac{12}{55}$$

22. (Section 8.2)

$$\binom{6}{5}\left(\frac{1}{6}\right)^5\left(\frac{5}{6}\right) = \frac{5}{7776}$$

23. (Section 8.2)

$$\binom{6}{4}\left(\frac{1}{6}\right)^4\left(\frac{5}{6}\right)^2 + \binom{6}{5}\left(\frac{1}{6}\right)^5\left(\frac{5}{6}\right) + \binom{6}{6}\left(\frac{1}{6}\right) = 0.04889$$

24. (Section 8.2)

$$\binom{10}{2}(.04)^2(.96)^8 = 0.05194$$

25. (Section 8.2)

$$\binom{10}{0}(.96)^{10} + \binom{10}{1}(.04)(.96)^9 + \binom{10}{2}(.04)(.96)^8 = 0.9938$$

26. Section 8.2)

$$\binom{10}{0}(.96)^{10} = 0.6648$$

27. Section 8.2)

$$\binom{10}{0}(.96)^{10} + \binom{10}{1}(.04)(.96)^9 + \binom{10}{2}(.04)^2(.96)^8 + \binom{10}{3}(.04)^3(.96)^7 = 0.9996$$

28. (Section 8.2)

$$1 - \left[\binom{15}{12}(.8)^{12}(.2)^3 + \binom{15}{13}(.8)^{13}(.2)^2 + \binom{15}{14}(.8)^{14}(.2) + \binom{15}{15}(.8)^{15}\right] = 1 - [0.6482]$$

$= 0.3518 = 35.18\%$

29. (Section 8.5)

Sample space = {2, 3, 4, 5, 6, 7, 8, 9, 10, 11, 12}
Probabilities = {1/36,1/18,1/12,1/9, 5/36, 1/6, 5/36, 1/9, 1/12, 1/18, 1/36}

30. (Section 8.5)

Sample space = {0, 1, 2, 3}
Probabilities = {1/8, 3/8, 3/8, 1/8}

31. (Section 8.1)

$P(A \mid P) = 0.8$

32. (Section 8.1)

$$P(A) = P(P) \bullet P(A \mid P) + P(Q) \bullet P(A \mid Q) + P(R) \bullet P(A \mid R)$$
$$= (.2)(.8) + (.7)(.4) + (.1)(.7) = .16 + .28 + .07 = 0.51$$

33. (Section 8.1)

$P(\bar{A} \mid Q) = 0.6$

34. (Section 8.2)

0.0446

35. (Section 8.1)

$$\frac{(.01)(.2)}{(.01)(.2) + (.02)(.4) + (.01)(.4)} = 0.1429$$

36. (Section 8.1)

$0.4 + 0.6 - 0.2 = 0.8$

37. (Section 8.2)

$$\binom{10}{9}\left(\frac{1}{6}\right)^9\left(\frac{5}{6}\right) = 0.00000083$$

38. (Section 8.2)

$$\binom{10}{9}\left(\frac{1}{6}\right)^9\left(\frac{5}{6}\right) + \left(\frac{1}{6}\right)^{10} = 0.00000084$$

CHAPTER 9

STATISTICS

For problems 1 – 5, the following data represent the number of car accidents per month in a small town over a two-year period.

Number of accidents

	Jan	Feb	Mar	Apr	May	Jun	Jul	Aug	Sep	Oct	Nov	Dec
Yr 1	169	163	170	165	165	169	168	172	170	172	171	165
Yr 2	170	168	177	164	173	166	173	176	177	172	170	172

1. Construct a frequency distribution based on the intervals 163 – 165, 166 – 168, 169 – 171, 172 – 174, 175 – 177.

2. Use the frequency distribution based on the intervals 163 – 165, 166 – 168, 169 – 171, 172 – 174, 175 – 177 to construct a histogram for the data.

3. Draw the frequency polygon for the data from a frequency distribution based on the intervals 163 – 165, 166 – 168, 169 – 171, 172 – 174, 175 – 177.

4. Find the cumulative (less than) frequencies for the data frequency distribution based on the intervals 163 – 165, 166 – 168, 169 – 171, 172 – 174, 175 – 177.

5. Draw a cumulative (less than) frequency distribution for the data frequency distribution based on the intervals 163 – 165, 166 – 168, 169 – 171, 172 – 174, 175 – 177.

For problems 6 – 11, use the following sample: 28, 30, 24, 30, 32, 40, 22, 25, 26, 34

6. Find the mean.

7. Find the median.

8. Find the mode.

9. Find the standard deviation.

10. Find the z-score for 30. (Assuming a normal distribution)

11. Find the range.

12. Find a z-score such that 10% of the area under the curve is to the left of z.

13. Find a z-score such that 96% of the area under the curve is to the left of z.

For problems 14 – 16, the mean of the set of data with the following grouped data is 27.

Class Interval	Frequency
20-22	11
23-25	9
26-28	12
29-31	15
32-34	8

14. Determine the standard deviation for this set of data.

15. Determine the first and third quartiles for this set of data.

16. Determine the centile point.

For problems 17 – 18, a probability distribution has an expected value of 28 and a standard deviation of 4. Use Chebyshev's Theorem to decide what is the least percent of the distribution described:

17. Between 20 and 36.

18. More than 22 and less than 34.

19. What is the smallest percentage of a sample that will be within 1.6 standard deviations of the mean?

For problems 20 – 22 using z-scores, what is the area under the standard normal curve:

20. To the left of 1.8?

21. To the right of 1.32?

22. Between –2.20 and 1.36?

23. Use the normal approximation to the binomial distribution to find the probability that at least 70 of 100 sick people will be cured by a new drug, when the probability is 0.75 that any one of them will be cured by the new drug.

24. Given a normal distribution with a mean of 8.2 and a standard deviation of 2.4, find the z-score for 9.1.

25. The time spent in a waiting line at a supermarket is known to be normally distributed with a mean of five minutes and a standard deviation of 30 seconds (0.5 minutes). Determine the probability that a randomly chosen customer will spend:
    (a) More than six minutes in the line.
    (b) Less than 5.5 minutes in the line.
    (c) Between 4.5 and 6.5 minutes in the line.
    (d) More than 4 minutes in the line.

26. For each of the following, give the mean, median, and mode.
    (a) 2, 3, 4, 5, 6
    (b) 2, 2, 2, 4, 5, 7, 7, 7, 8, 9
    (c) 1, 2, 4, 4, 6, 11

27. Which of the following is not a measure of central tendency?
    (a) mode
    (b) standard deviation
    (c) mean
    (d) median

28. **Capital Gains Tax.** The following tax rates in percents is from the U.S. Chamber of Commerce. Plot using a bar graph.

YEAR	EFFECTIVE RATE	YEAR	EFFECTIVE RATE
1960	25.0	1976	49.1
1968	26.9	1979	28.0
1969	27.5	1981	20.0
1970	32.2	1987	28.0
1971	34.4	1988	33.0
1972	45.5	1991	28.0

29. **Consumer Price Indexes.** The following list of percent changes for all items is from the Bureau of Labor Statistics; U.S. Dept. of Labor. Plot using a bar graph.

YEAR	INFLATION RATE	YEAR	INFLATION RATE
1978	7.6	1985	3.6
1979	11.3	1986	1.9
1980	13.5	1987	3.6
1981	10.3	1988	4.1
1982	6.2	1989	4.8
1983	3.2	1990	5.4

30. **Foreign Direct Investment In U.S.** The following figures (in billions of U.S. dollars), for the year 1990, are from the Bureau of Economic Analysis; U.S. Dept. of Commerce. Plot using a pie chart.

All Countries	403.7
Canada	27.7
Germany	27.8
Japan	108.1
Netherlands	83.5
Switzerland	64.5
U.K.	17.5

31. **Scholastic Aptitude Test (SAT) Scores.** The following figures are from the College Entrance Examination Board for combined verbal and math. Plot a bar graph.

YEAR	1970	1975	1980	1982	1984	1985	1987	1988	1989	1990	1991
SCORE	948	906	890	893	897	906	906	904	903	900	896

32. **Municipal Solid Waste Generation.** The following figures (in millions of tons) for 1998 are from the Environmental Protection Agency. Plot a pie chart.

SOURCE	AMOUNT
Paper	71.8
Yard Waste	31.6
Metals	15.3
Plastics	14.4
Food Wastes	13.2
Glass	12.5
Rubber, Leather, Textile, Wood, All Other	20.8

For questions 33 – 34, consider a test given to a large population of students, The result can be considered normally distributed with a mean score of 1,300 and a standard deviation of 100.

33. Find the probability a test picked up from a stack will have a score between 1,200 and 1,500.

34. Find the probability a test picked up from a stack will have a score of at least 1,500.

For questions 35 – 37, consider the following data set: 1, 2, 2, 4, 4, 4, 8, 8, 10, 10, 10.

35. Find the mean.

36. Find the median.

37. Find the mode(s).

SOLUTIONS

1. (Section 9.2)

Interval	Frequency
163-165	5
166-168	3
169-171	7
172-174	6
175-177	3

2. (Section 9.2)

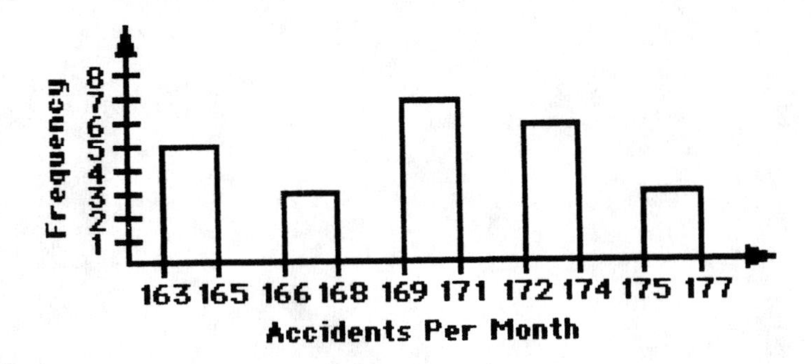

3. (Section 9.2)

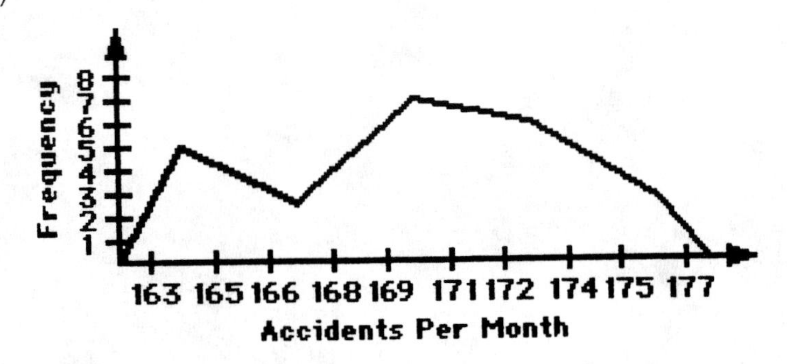

4. (Section 9.2)

Interval	Frequency	Cumulative Frequency
163-165	5	5
166-168	3	8
169-171	7	15
172-174	6	21
175-177	3	24

5. (Section 9.2)

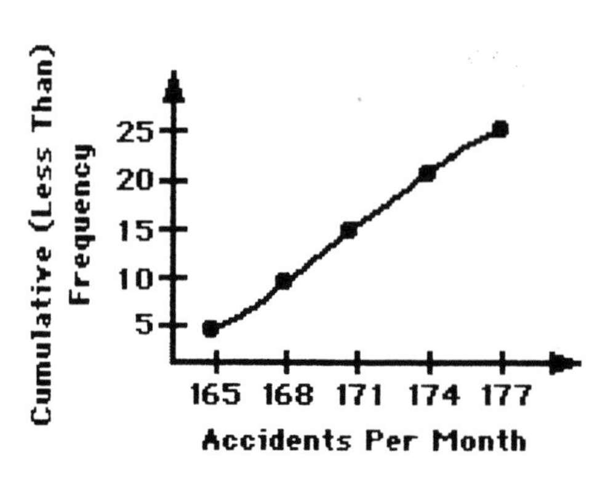

6. (Section 9.3)
29.1

7. (Section 9.3)
29

8. (Section 9.3)
30

9. (Section 9.4)
5.069

10. (Section 9.5)
$$\frac{30 - 29.1}{5.069} = .1775$$

11. (Section 9.2)
40 – 22 = 18

12. (Section 9.5)
–1.282

13. (Section 9.5)
1.751

14. (Section 9.4)
$$\sigma = \sqrt{\frac{11(21-27)^2 + 9(24-27)^2 + 12(27-27)^2 + 15(30-27)^2 + 8(33-27)^2}{55}}$$
$$= 4.05$$

15. (Section 9.3)

$$C_{25} = 22.5 + \left(\frac{3.25}{9}\right) \bullet 3 = 23.58$$

$$C_{75} = 28.5 + \left(\frac{9.75}{15}\right) \bullet 3 = 30.45$$

16. (Section 9.3)

$$C_{50} = 25.5 + \left(\frac{8}{12}\right) \cdot 3 = 27.5$$

17. (Section 9.4)

$k = 36 - \bar{X} = 36 - 28 = 8;$

Probability at least $1 - \frac{\sigma^2}{k^2} = 1 - \frac{4^2}{8^2} = \frac{3}{4} = 75\%$

18. (Section 9.4)

$k = 34 - \bar{X} = 6; \left(1 - \frac{\sigma^2}{k^2}\right) = \left(1 - \frac{4^2}{(1.5 \cdot 4)^2}\right) = .55\bar{5}$

Probability more than 22 and less than 34:

$.55\bar{5}$

19. (Section 9.4)
$k = 1.6\sigma = 6.4;$

Probability at least $1 - \frac{\sigma^2}{k^2} = 1 - \frac{4^2}{6.4^2} = 60.9\%$

20. (Section 9.5)
0.9641

21. (Section 9.5)
1 – .9066 = .0934

22. (Section 9.5)
.9131 – .0139 = .8992

23. (Section 9.5)

$\bar{X} = np = 100(.75) = 75; \sigma = \sqrt{npq} = \sqrt{100(.75)(.25)} = 4.33$

$$P(x \geq 70) = P\left(\frac{x - 75}{4.33} \geq \frac{70 - 75}{4.33}\right) = P(z \geq -1.155) = 1 - .1240 = .8760$$

24. (Section 9.5)

$\frac{9.1 - 8.2}{2.4} = .375$

25. (Section 9.5)

(a) $P(x > 6) = P\left(\frac{x - 5}{5} > \frac{6 - 5}{.5}\right) = P(z > 2) = 1 - .9772 = .0228$

(b) $P(x < 5.5) = P(z < 1) = .8413$
(c) $P(4.5 < x < 6.5) = P(-1 < z < 3) = .9987 - .1587 = .8400$

(d) $P(x > 4) = P(z > -2) = 1 - .0228 = .9772$

26. (Section 9.3)
(a) mean = 4
median = 4
mode = no mode

(b) mean = 5.3
median = 6 (average of 2 middle scores)
mode = 2, 7 (bimodal)

(c) mean = 4 2/3
median = 4
mode = 4

27. (Section 9.3)
The standard deviation is a measure of dispersion and *not* a measure of central tendency. Thus "B" is the correct answer.

28. (Section 9.1)

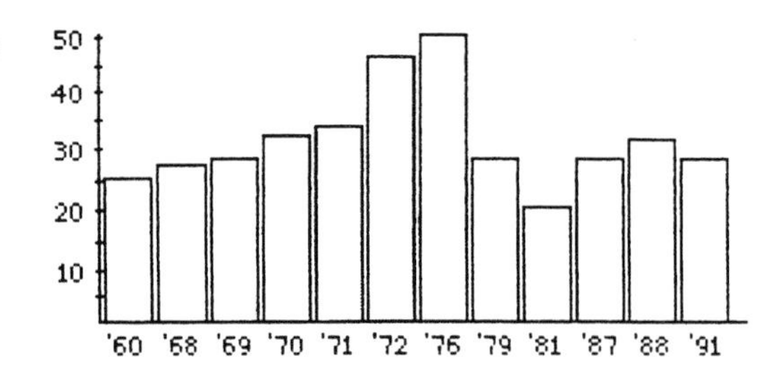

29. (Section 9.1)

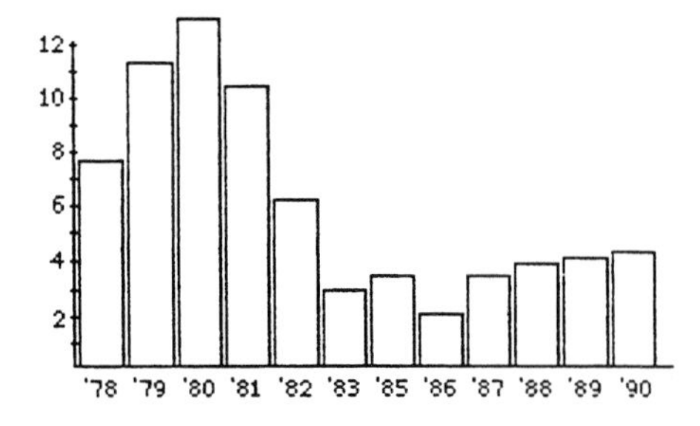

30. (Section 9.1)

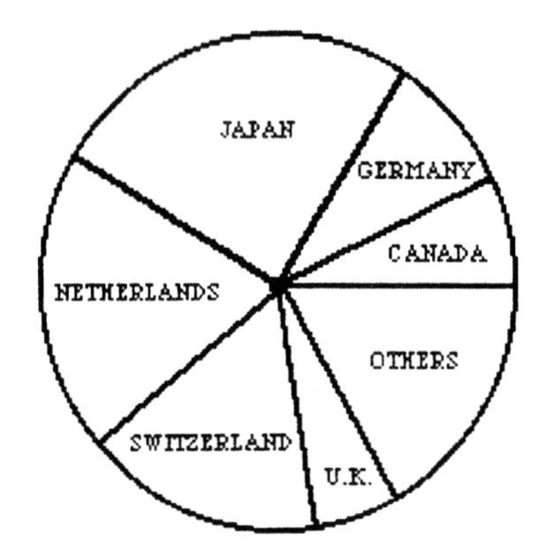

31. (Section 9.1)

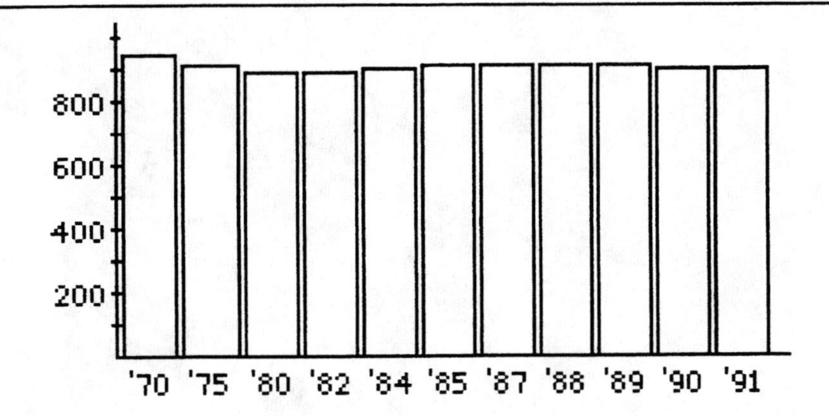

32. (Section 9.1)

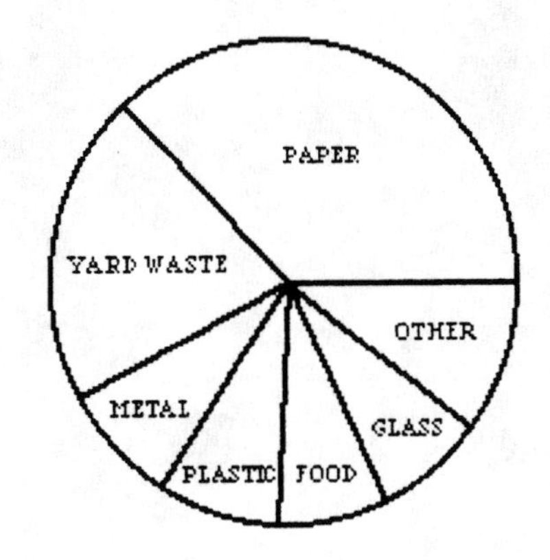

33. (Section 9.5)
0.9772 – 0.1597 = 0.8175

34. (Section 9.5)
1 - 0.9772 = 0.0228

35. (Section 9.3)

$$\frac{1+2+2+4+4+4+8+8+10+10+10}{10} = 5.4$$

36. (Section 9.3)
4

37. (Section 9.3)
4, 10

CHAPTER 10

MARKOV CHAINS; GAMES

In Problems 1 – 6, tell whether or not the matrix could be a transition marix for a Markov chain. If it cannot be a transition matrix, state why.

1. $\begin{bmatrix} -1/2 & 3/2 \\ 1/3 & 2/3 \end{bmatrix}$

2. $\begin{bmatrix} 1/2 & 1/3 & 1/4 \\ 0 & 1 & 0 \\ 1/2 & 0 & 1/2 \end{bmatrix}$

3. $\begin{bmatrix} 1/2 & 1/2 \\ 1 & 0 \\ 1/4 & 3/4 \end{bmatrix}$

4 $\begin{bmatrix} 2 & .8 \\ 1 & 0 \end{bmatrix}$

5 $\begin{bmatrix} .3 & .8 & -.1 \\ .7 & .1 & .2 \\ -3 & 0 & .7 \end{bmatrix}$

6. $\begin{bmatrix} 1 & 0 & 0 \\ 0 & 1 & 0 \\ 0 & 0 & 1 \end{bmatrix}$

7. Consider the Markov chain:

$$\begin{bmatrix} .6 & .2 & .2 \\ .4 & .1 & .5 \\ .7 & .2 & .1 \end{bmatrix}$$

If the initial distribution is [.25 .25 .5] , what is the probability distribution in the next observation?

In problems 8 – 11, use the transition matrix P, together with the initial distribution $A^{(o)}$, to find the distribution after two stages of the experiment. Then, predict the long-range distribution.

8. $A^{(o)} = [\,.3 \quad .7\,]$ $\quad P = \begin{bmatrix} .4 & .6 \\ .5 & .5 \end{bmatrix}$

9. $A^{(o)} = [\,.2 \quad .8\,]$ $\quad P = \begin{bmatrix} .9 & .1 \\ .4 & .6 \end{bmatrix}$

10. $A^{(o)} = [\,.2 \quad .5 \quad .5\,]$ $\quad P = \begin{bmatrix} .2 & .2 & .6 \\ .3 & .3 & .4 \\ .1 & 0 & .9 \end{bmatrix}$

11. $A^{(0)} = [\,.1 \quad 0 \quad .9\,]$ $\quad P = \begin{bmatrix} .2 & .2 & .6 \\ .3 & .3 & .4 \\ .1 & 0 & .9 \end{bmatrix}$

12. Let the population of a county be classified according to income

    State 1: poor
    State 2: middle income
    State 3: rich

    Over a 25 year period (one generation), we have the following data:

    Of the poor, 19% become middle income and 1% rich.
    Of the middle income, 15% become poor and 10% rich.
    Of the rich, 5% become poor and 30% middle income.

    a. Give the transition matrix for the above data.
    b. For the above Markov chain, find the transition matrix for the next 25 years (i.e., after 50 years).
    c. Is the matrix found in part "b" a transition matrix? Why or why not?

13. Find a, b, and c values to make this a transition matrix for a Markov chain.

$$\begin{bmatrix} .1 & a & .3 \\ b & .2 & .7 \\ 0 & c & 0 \end{bmatrix}$$

14. Find the fixed probability vector "t" of the transition matrix

$$\begin{bmatrix} 1/3 & 2/3 \\ 1/2 & 1/2 \end{bmatrix}$$

For problems 15 – 22, is the matrix regular? State why.

15. $\begin{bmatrix} 1/2 & 1/2 \\ 1/3 & 2/3 \end{bmatrix}$

16. $\begin{bmatrix} 0 & 1 \\ 1/4 & -3/4 \end{bmatrix}$

17. $\begin{bmatrix} .2 & .4 & .4 \\ .1 & .5 & .4 \\ .3 & .1 & .8 \end{bmatrix}$

18. $\begin{bmatrix} .5 & 0 & 0 & .5 \\ .5 & .5 & 0 & 0 \\ 0 & .5 & .5 & 0 \\ 0 & 0 & .5 & .5 \end{bmatrix}$

19. $\begin{bmatrix} .3 & .4 & .1 & .2 \\ .2 & .2 & .2 & .4 \\ 0 & 0 & 1 & 0 \\ .5 & .1 & .2 & .2 \end{bmatrix}$

20. $\begin{bmatrix} a & 0 & 0 \\ b & c & d \\ e & f & g \end{bmatrix}$

21 $\begin{bmatrix} 0 & a & 0 \\ b & c & d \\ 0 & e & 0 \end{bmatrix}$

22. $\begin{bmatrix} 0 & a & 0 & 0 \\ b & 0 & 0 & 0 \\ c & d & e & f \\ g & h & i & j \end{bmatrix}$

For problems 23 – 25:

$$\text{Let } T = \begin{bmatrix} .3 & .4 & .3 \\ .7 & .2 & .1 \\ .4 & .1 & .5 \end{bmatrix}$$

be the transition matrix for a Markov chain and let P = [.3 .2 .5] be the initial population distribution vector.

23. Find the proportion of the state 2 population that is in state 3 after 2 time periods.

24. Find the proportion of the state 3 population that is in state 2 after 2 time periods.

25. Find the proportion of the total population that is in state 3 after 2 time periods.

26. The inhabitants of a vegetarian prone community have the following rules:
   a. A person who eats no meat one day will flip a fair coin and eat meat on the next day if and only if a tail appears.
   b. No one can eat meat 2 days in a row.

   Determine if this Markov chain is regular and if so, find its probability vector.

For problems 27 – 36, mark the question as being True or False.

27. All entries in a transition matrix are non-negative.

28. Every matrix whose entries are all non-negative is a transition matrix.

29. The sum of all the entries in an "n x n" transition matrix is 1.

30. The sum of all the entries in an "n x n" transition matrix is "n".

31. If a transition matrix contains no zero entries, it is regular.

32. If a transition matrix is regular, it contains no nonzero entries.

33. Every power of a transition matrix is again a transition matrix.

34. If a transition matrix is regular, its square has equal row vectors.

35. If a transition matrix T is regular, there exists a unique vector S such that ST = S.

36. If a transition matrix T is regular, there exists a unique population distribution vector S such that ST = S.

For problems 37 – 39, find the expected payoff for each game

$$\begin{bmatrix} 2 & 0 \\ 1 & 3 \end{bmatrix}$$

for the given strategy.

37. $P = [\ 1/2 \quad 1/2\ ]$ $\qquad Q = \begin{bmatrix} 1/2 \\ 1/2 \end{bmatrix}$

38. $P = [\ 1/4 \quad 3/4\ ]$ $\qquad Q = \begin{bmatrix} 3/4 \\ 1/4 \end{bmatrix}$

39. $P = [\ 1 \quad 0\ ]$ $\qquad Q = \begin{bmatrix} 0 \\ 1 \end{bmatrix}$

40. Using the game matrix: $\begin{bmatrix} 2 & 0 \\ 1 & 3 \end{bmatrix}$

   a. What is the optimal strategy for player I?
   b. What is the optimal strategy for player II?
   c. Find the value of the game when the optimal strategies are used.
   d. Which player does the game favor?

41. If the game matrix is $A = \begin{bmatrix} 1 & -3 & 0 \\ 0 & 2 & -1 \end{bmatrix}$

   find the expected payoff if $P = [\ 1/2 \quad 1/2\ ]$ and $Q = \begin{bmatrix} 1/3 \\ 1/2 \\ 1/6 \end{bmatrix}$

For problems 42 – 49 determine if the 2 person zero-sum game is strictly determined. If it is, find the value of the game. All entries are winnings of Player I who plays rows.

42. $\begin{bmatrix} -2 & 4 \\ -5 & 5 \end{bmatrix}$

43. $\begin{bmatrix} 3 & 1 \\ 4 & 2 \end{bmatrix}$

44. $\begin{bmatrix} 3 & 0 \\ 0 & -2 \end{bmatrix}$

45 $\begin{bmatrix} -5 & -2 \\ 0 & 0 \end{bmatrix}$

46. $\begin{bmatrix} 1 & 0 & -2 \\ 4 & 5 & 0 \\ 0 & 2 & 4 \end{bmatrix}$

47. $\begin{bmatrix} 1 & 3 & 7 \\ 3 & 5 & 0 \\ 2 & 0 & -3 \end{bmatrix}$

48. $\begin{bmatrix} 2 & 4 & 6 & 8 \\ 1 & 2 & 4 & -5 \\ 0 & -5 & 6 & 9 \end{bmatrix}$

49. $\begin{bmatrix} 1 & -1 & 6 & 9 \\ 5 & -4 & 4 & 0 \\ 0 & -3 & 1 & 0 \end{bmatrix}$

50. Are any of the following game matrices from problems 42 – 49 fair? If so, which one(s)?

a. $\begin{bmatrix} -2 & 4 \\ -5 & 5 \end{bmatrix}$ b. $\begin{bmatrix} 3 & 1 \\ 4 & 2 \end{bmatrix}$ c. $\begin{bmatrix} 3 & 0 \\ 0 & -2 \end{bmatrix}$ d. $\begin{bmatrix} -5 & -2 \\ 0 & 0 \end{bmatrix}$

e. $\begin{bmatrix} 1 & 0 & -2 \\ 4 & 5 & 0 \\ 0 & 2 & 4 \end{bmatrix}$ f. $\begin{bmatrix} 1 & 3 & 7 \\ 3 & 5 & 0 \\ 2 & 0 & -3 \end{bmatrix}$

g. $\begin{bmatrix} 2 & 4 & 6 & 8 \\ 1 & 2 & 4 & -5 \\ 0 & -5 & 6 & 9 \end{bmatrix}$ h. $\begin{bmatrix} 1 & -1 & 6 & 9 \\ 5 & -4 & 4 & 0 \\ 0 & -3 & 1 & 0 \end{bmatrix}$

In problems 51 – 52, for what values of x is the matrix strictly determined?

51. $\begin{bmatrix} x & 1 \\ 2 & -3 \end{bmatrix}$

52. $\begin{bmatrix} x & 8 & 3 \\ 0 & x & -9 \\ -2 & 2 & x \end{bmatrix}$

53. What values of c for the given game matrix would result in a saddle point of 2?

$$\begin{bmatrix} 2 & 4 & c \\ 0 & -5 & 6 \\ 1 & 3 & -2 \end{bmatrix}$$

54. In a game, Player 1 picks heads or tails and Player 2 tries to guess the choice. Player 1 will pay Player 2 \$4 if both choose heads. Player 1 will pay Player 2 \$3 if both choose tails. If Player 2 guesses incorrectly, he will pay Player 1 \$6.

a. Give the pay-off matrix.

b. Is the above game strictly determined?

55. A professor decides on a quiz tomorrow according to whether or not she has a quiz today using the following procedure: If there is a quiz today, she writes a number from 1 to 5 on a piece of paper and has a student guess it at the end of class. The professor gives a quiz tomorrow if the student correctly guesses her number, otherwise there is no quiz tomorrow. If no quiz is given today, then the professor flips a penny and has a quiz tomorrow if it shows tails (no quiz if it shows heads).
    a. Write the transition matrix for this process.
    b. If there is a quiz today, what is the expected number of days until the next quiz?

56. a. Rearrange the transition matrix and find the substructures S and Q.

$$\begin{bmatrix} 0 & 1 & 0 & 0 \\ 0 & 0 & 0 & 1 \\ 0 & .1 & .9 & 0 \\ .7 & .3 & 0 & 0 \end{bmatrix}$$

    b. Find T (the fundamental matrix of Markov chain).
    c. Find T • S.
    d. Starting in state 2, find the probability of reaching state 1.

In problems 57 – 60, find the fundamental matrix T and T • S for each absorbing matrix.

57. $$\begin{bmatrix} 1 & 0 & 0 \\ 0 & 1 & 0 \\ .3 & .2 & .5 \end{bmatrix}$$

58. $$\begin{bmatrix} .2 & .7 & .1 \\ 0 & 1 & 0 \\ 0 & 0 & 1 \end{bmatrix}$$

59. $$\begin{bmatrix} 1 & 0 & 0 & 0 \\ .1 & .2 & .3 & .3 \\ .2 & .5 & .1 & .2 \\ 0 & 0 & 0 & 1 \end{bmatrix}$$

60. $$\begin{bmatrix} 1/6 & 1/3 & 1/3 & 1/6 \\ 0 & 1 & 0 & 0 \\ 0 & 0 & 1 & 0 \\ 1/8 & 1/4 & 1/4 & 3/8 \end{bmatrix}$$

In problems 61 – 69, tell whether the matrix is absorbing.

61. $$\begin{bmatrix} .6 & .2 & .2 & 0 \\ .1 & 0 & .5 & .4 \\ 0 & 1 & 0 & 0 \\ .1 & 0 & .3 & .6 \end{bmatrix}$$

62. $$\begin{bmatrix} 1 & 0 & 0 \\ 0 & 1/2 & 1/2 \\ 0 & 1/4 & 3/4 \end{bmatrix}$$

63. $\begin{bmatrix} 0 & 0 & 1 \\ 1/3 & 1/3 & 1/3 \\ 1 & 0 & 0 \end{bmatrix}$

64. $\begin{bmatrix} 0 & 1 \\ 3/4 & 1/4 \end{bmatrix}$

65. $\begin{bmatrix} 1/4 & 1/2 & 0 & 1/4 \\ 0 & 1/3 & 1/3 & 1/3 \\ 0 & 0 & 1 & 0 \\ 0 & 1/2 & 0 & 1/2 \end{bmatrix}$

66. $\begin{bmatrix} .3 & .7 & 0 \\ .5 & .5 & 0 \\ 0 & 0 & 1 \end{bmatrix}$

67. $\begin{bmatrix} .1 & .4 & .5 \\ 0 & 1 & 0 \\ 1 & 0 & 0 \end{bmatrix}$

68. $\begin{bmatrix} .2 & .1 & .2 & .5 \\ 0 & 0 & 1 & 0 \\ 0 & 1 & 0 & 0 \\ .1 & .1 & .1 & .7 \end{bmatrix}$

69. $\begin{bmatrix} .3 & .2 & 0 & .5 \\ 0 & .3 & .4 & .3 \\ 0 & 0 & 1 & 0 \\ 0 & .5 & 0 & .5 \end{bmatrix}$

70. Below is the fundamental matrix T for a certain absorbing Markov Chain:

	\$1	\$2	\$3
\$1	.5	.8	1.5
\$2	1.2	.6	2.3
\$3	.3	1.8	2.1

a. What is the expected number of times a person starting with \$1 will have \$3?
b. What is the expected number of times a person starting with \$1 will have \$2?
c. If a player starts with \$3, how many games can he expect before absorption?

For problems 71 – 73, in a certain town 60% of the children read children's books, and 40% do not. Suppose that the following transition matrix indicates the change from one year to the next for readers and non readers:

	Reader	Non Reader
Reader	.8	.2
Non Reader	.3	.7

71. Find the percentage of readers next year.

72. Find the percentage of readers two years from now.

73. What will the long-range percentage be?

For problems 74 – 76, in a certain town 70% of the people drink and 30% do not. An anti-drinking campaign had the following effect on people, before and after the campaign:

	Drinker	Non Drinker
Drinker	.9	.1
Non Drinker	.2	.8

74. What will be the effect on people after one campaign?

75. What will be the effect on people after two campaigns?

76. What will the distribution be in the long run?

For problems 77 – 86, a study of the hair color of females and their offspring resulted in the following transition matrix:

		Offspring		
		Blonde	Brunette	Redhead
Present Individual	Blonde	.5	.3	.2
	Brunette	.3	.6	.1
	Redhead	.2	.6	.2

77. Find the probability that a woman with blonde hair will have a blonde child.

78. Find the probability that a woman with blonde hair will have a blonde grandchild.

79. Find the probability that a woman with blonde hair will have a blonde great-grandchild.

80. Find the probability that a brunette woman will have a brunette child.

81. Find the probability that a brunette woman will have a redhead grandchild.

82. Find the probability that a brunette woman will have a blonde great-grandchild.

83. If the population of women is now 40% brunettes, 50% blondes, and the rest redheads, find the distribution after one generation.

84. If the population of women is now 40% brunettes, 50% blondes, and the rest redheads, find the distribution after two generations.

85. If the population of women is now 40% brunettes, 50% blondes, and the rest redheads, find the distribution after three generations.

86. Find the long-run distribution of hair color.

For problems 87 – 90, three fighter planes, A, B, and C, are engaged in a three-way battle. Fighter plane A has probability 1/4 of destroying its target, B has probability 1/2 of destroying its target, and C has probability 1/4 of destroying its target. The planes fire at the same time and each fires at the strongest opponent not yet destroyed. Using as states the surviving airplanes at any round, set up a Markov chain and answer the following:

87. How many states are in this chain?

88. How many states are absorbing?

89. Find the expected number of rounds fired.

90. Find the probability that plane A survives.

Use the following information for problems 91 – 94

A professor has a habit of giving quizzes. If there is a quiz on any given day, there is a 0.2 probability of a quiz on the next day. If there is no quiz on any day, there is a probability of 0.6 there will be a quiz on the next day.

91. Write the transition matrix.

92. If the professor advertises that on the first day of class there is a 0.5 probability of having a quiz, what is the probability there will be a quiz on day two?

93. If the professor advertises that on the first day of class there is a 0.5 probability of having a quiz, what is the probability there will be a quiz on day three?

94. If the professor advertises that on the first day of class there is a 0.5 probability of having a quiz, what is the probability there will be a quiz on the last day of class?

SOLUTIONS

1. (Section 10.1)
   No, because you have a negative entry.

2. (Section 10.1)
   No, because the entries in row 1 do not total 1.

3. (Section 10.1)
   No, because the matrix is not square.

4. (Section 10.1)
   No, because row 1 does not sum to 1.

5. (Section 10.1)
   No, because rows 1 & 3 have negative entries.

6. (Section 10.1)
Yes

7. (Section 10.1)

$$[.25 \quad .25 \quad .5]\begin{bmatrix} .6 & .2 & .2 \\ .4 & .1 & .5 \\ .7 & .2 & .1 \end{bmatrix} = [.6 \quad .175 \quad .225]$$

8. (Section 10.2)

$$A^{(1)} = A^{o}P = [\,.3 \quad .7\,]\begin{bmatrix} .4 & .6 \\ .5 & .5 \end{bmatrix} = [.47 \quad .53] = A^{1}$$

$$A^{2} = A^{1}P = [.47 \quad .53]\begin{bmatrix} .4 & .6 \\ .5 & .5 \end{bmatrix} = [.453 \quad .547]$$

Long range, solve:

$$[\,x \quad 1-x\,]\begin{bmatrix} .4 & .6 \\ .5 & .5 \end{bmatrix} = [\,x \quad 1-x\,]$$

Answer: [.455 .545]

9. (Section 10.2)

$$A^{1} = A^{o}P = [\,.2 \quad .8\,]\begin{bmatrix} .9 & .1 \\ .4 & .6 \end{bmatrix} = [.5 \quad .5]$$

$$A^{2} = A^{1}P = [\,.5 \quad .5\,]\begin{bmatrix} .9 & .1 \\ .4 & .6 \end{bmatrix} = [\,.65 \quad .35\,]$$

Long range: [.8 .2]

10. (Section 10.2)

$$A^{1} = A^{o}P = [.2 \quad .3 \quad .5]\begin{bmatrix} .2 & .2 & .6 \\ .3 & .3 & .4 \\ .1 & 0 & .9 \end{bmatrix} = [.18 \quad .13 \quad .69]$$

$$A^{2} = A^{1}P = [.18 \quad .13 \quad .69]\begin{bmatrix} .2 & .2 & .6 \\ .3 & .3 & .4 \\ .1 & 0 & .9 \end{bmatrix} = [.144 \quad .075 \quad .781]$$

Long range: $x + y + z = 1 \Rightarrow z = 1 - x - y$

$$[\,x \quad y \quad 1-x-y\,]\begin{bmatrix} .2 & .2 & .6 \\ .3 & .3 & .4 \\ .1 & 0 & .9 \end{bmatrix} = [\,x \quad y \quad 1-x-y\,]$$

$$\begin{aligned} 2x + .3y + .1 - .1x - .1y &= x \\ .2x + .3y + 0(1 - x - y) &= y \end{aligned}$$

$.6x + .4y + .9 - .9y - .9x = 1 - x - y$

Solving simultaneously: [.1186 .0339 .8475]

11. (Section 10.2)

$$A^1 = A^oP = [.1 \quad 0 \quad .9] \begin{bmatrix} .2 & .2 & .6 \\ .3 & .3 & .4 \\ .1 & 0 & .9 \end{bmatrix} = [.11 \quad .02 \quad .87]$$

$$A^2 = A^1P = [.11 \quad .02 \quad .87] \begin{bmatrix} .2 & .2 & .6 \\ .3 & .3 & .4 \\ 1 & 0 & .9 \end{bmatrix} = [.115 \quad .028 \quad .857]$$

Long range: $x + y + z = 1 \Rightarrow z = 1 - x - y$

$$[x \quad y \quad 1-x-y] \begin{bmatrix} .2 & .2 & .6 \\ .3 & .3 & .4 \\ .1 & 0 & .9 \end{bmatrix} = [x \quad y \quad 1-x-y]$$

$$\begin{aligned} 2x + .3y + .1 - .1x - .1y &= x \\ .2x + .3y + 0(1 - x - y) &= y \\ .6x + .4y + .9 - .9y - .9x &= 1 - x - y \end{aligned}$$

Solving simultaneously: [.1186 .0339 .8475]

12. a. (Section 10.1)

$$\begin{bmatrix} .80 & .19 & .01 \\ .15 & .75 & .10 \\ .05 & .30 & .65 \end{bmatrix} = T$$

b. (Section 10.2)

$$\begin{bmatrix} .6690 & .2975 & .0335 \\ .2375 & .6210 & .1415 \\ .1175 & .4295 & .4530 \end{bmatrix} = T^2$$

c. (Section 10.1)
Yes, because the sum of the entries in each row is 1.

13. (Section 10.1)

$.1 + a + .3 = 1$, $a = .6$ $b + .2 + .7 = 1$, $b = .1$ $0 + c + 0 = 1$, $c = 1$

14. (Section 10.2)

$$[t_1 \quad t_2] \begin{bmatrix} 1/3 & 2/3 \\ 1/2 & 1/2 \end{bmatrix} = [t_1 \quad t_2] \text{ and } t_1 + t_2 = 1.$$

Therefore, $[\,1/3\, t_1 + 1/2\, t_2 \quad 2/3\, t_1 + 1/2\, t_2\,] = [\,t_1 \quad t_2\,]$

which implies that

$$\begin{cases} 1/3\, t_1 + 1/2\, t_2 = t_1 \\ 2/3\, t_1 + 1/2\, t_2 = t_2 \end{cases}$$

Subtracting the first equation from the second one yields

$1/3\, t_1 = t_2 - t_1$ or $4/3\, t_1 - t_2 = 0.$

Combining this result with the condition that the sum of the probability vector is 1, we get

$$\begin{cases} 4/3\, t_1 - t_2 = 0 \\ t_1 + t_2 = 1 \end{cases}$$

Adding these two equations leads to $7/3\, t_1 = 1$. Hence $t_1 = 3/7$ and $t_2 = 4/7$. So, our solution is [3/7 4/7].

15. (Section 10.2)

$$T^2 = \begin{bmatrix} 5/12 & 7/12 \\ 7/18 & 11/18 \end{bmatrix}$$

Since T^2 has only positive entries, T (the given matrix) is a regular transition matrix.

16. (Section 10.2)
The matrix is not a transition matrix because it has a negative entry. Since it is not a transition matrix, it cannot be regular.

17. (Section 10.2)
The entries in row 3 do not add up to 1. Thus, it is not a transition matrix and cannot be regular.

18. (Section 10.2)

$$T^3 = \begin{bmatrix} .125 & .125 & .375 & .375 \\ .375 & .125 & .125 & .375 \\ .375 & .375 & .125 & .125 \\ .125 & .375 & .375 & .125 \end{bmatrix}$$

Since T^3 has all positive entries, the given matrix T is regular.

19. (Section 10.2)
Let the given matrix be T. T, T^2, T^3, T^4, etc. all have a third row of 0 0 1 0 and could never be regular.

20. (Section 10.2)
If T is the given transition matrix, T^n where n is any positive integer will always be in the form

$$\begin{bmatrix} a^n & 0 & 0 \\ x & x & x \\ x & x & x \end{bmatrix}$$

where the x is some positive integer. Hence, T is not regular.

21. (Section 10.2)
If T is the transition matrix, then

$$T^2 = \begin{bmatrix} ab & ac & ad \\ bc & ab + c^2 + de & cd \\ be & ce & de \end{bmatrix}$$

which has all positive entries and so T must be regular.

22. (Section 10.2)
T^n is always of the form
$$\begin{bmatrix} 0 & a^m b^{n-m} & 0 & 0 \\ a^{n-m} b^m & 0 & 0 & 0 \\ x & x & x & x \\ x & x & x & x \end{bmatrix}$$

where the x's represent positive numbers and n is odd. If n is even, then T^n is always of the form
$$\begin{bmatrix} a^{N/2} b^{N/2} & 0 & 0 & 0 \\ 0 & a^{N/2} b^{N/2} & 0 & 0 \\ x & x & x & x \\ x & x & x & x \end{bmatrix}$$

Since T^N oscillates between these two forms, it will never be regular.

23. (Section 10.2)

$$T^2 = \begin{bmatrix} .49 & .23 & .28 \\ .39 & .23 & .28 \\ .39 & .23 & .38 \end{bmatrix}$$

So, after 2 time periods, the proportion of State 2 population in State 3 = a_{23} = .28

24. (Section 10.2)

$$T^2 = \begin{bmatrix} .49 & .23 & .28 \\ .39 & .23 & .28 \\ .39 & .23 & .38 \end{bmatrix}$$ a_{32} in this T^2 matrix is .23

25. (Section 10.2)

PT^2 = [.42 .25 .33]
So, .33 is the proportion of the total population in State 3 after 2 time periods.

26. (Section 10.2)
$$\begin{bmatrix} 1/2 & 1/2 \\ 1 & 0 \end{bmatrix} = T$$

Since T^2 has all positive entries, T is regular. Solving for the probability vector:

$$[t_1 \quad t_2] \begin{bmatrix} 1/2 & 1/2 \\ 1 & 0 \end{bmatrix} = [t_1 \quad t_2]$$

yields $\begin{cases} 1/2\, t_1 + t_2 = t_1 \\ 1/2\, t_1 = t_2 \end{cases}$

Using the second equation and the fact that → $\begin{cases} 1/2\, t_1 - t_2 = 0 \\ t_1 + t_2 = 1 \end{cases}$

results in $3/2\, t_1 = 1$. So, $t_1 = 2/3$, $t_2 = 1/3$. Thus the probability vector is [2/3 1/3].

27. (Section 10.1)
True

28. (Section 10.1)
False

29. (Section 10.1)
False

30. (Section 10.1)
True

31. (Section 10.2)
True

32. (Section 10.2)
False

33. Section 10.1)
True

34. (Section 10.2)
False

35. (Section 10.2)
False

36. (Section 10.2)
True

37. (Section 10.5)
Expected Payoff = PAQ = [1.5] = $1.50

38. (Section 10.5)
Expected Payoff = PAQ = [1.5] = $1.50

39. (Section 10.5)
Expected Payoff = PAQ = [0] = $0

40. a. (Section 10.6)

Let $[\,p_1 \quad p_2\,] = P$

$$p_1 = \frac{3-1}{4} = 1/2 \text{ and } p_2 = \frac{2-0}{4} = 1/2$$

So, $P = [\ 1/2 \quad 1/2\ ]$ which means that Player 1 should select the first row 1/2 of the time and the second row 1/2 of the time.

b. (Section 10.6)
For Player 2, $Q = [q_1 \quad q_2]$

$$q_1 = \frac{3-0}{4} \text{ and } q_2 = \frac{2-1}{4}$$

So, $Q = \begin{bmatrix} 3/4 \\ 1/4 \end{bmatrix}$

Hence, Player 2 would do the best if he selected column 1 75% of the time and column 2 25% of the time.

c. (Section 10.5)

$$\text{Expected Payoff} = PAQ = \frac{6-0}{4} = \frac{6}{4} = \$1.50$$

d. (Section 10.5)
The game favors Player 1.

41. (Section 10.5)

$PAQ = -1/6 = -.1\overline{6}$ which is approximately –$.17

42. (Section 10.4)
Yes, strictly determined because –2 is the smallest number in row 1 and the largest number in column 1. Thus, the value of the game is –2.

43. (Section 10.4)
Yes, strictly determined and game value is 2.

44. (Section 10.4)
Yes, strictly determined and game value is 0.

45. (Section 10.4)
Yes, strictly determined and game value is 0.

46. (Section 10.4)
Yes, strictly determined and game value is 4.

47. (Section 10.4)
Not strictly determined.

48. (Section 10.4)
Yes, strictly determined and game value is 2.

49. (Section 10.4)
Yes, strictly determined and game value is –1.

50. (Section 10.4)
Yes, the game matrices in c and d are fair since its game value is zero.

51. (Section 10.4)
If $x < 1$, then x will be the smallest in the row and x needs to be greater than 2 to be the largest in the column. Therefore $2 \leq x \leq 1$. If $x \geq 2$, $a_{12} = 1$ is a saddle point.

52. (Section 10.4)
Combining the conditions for x to be a saddle point in each row gives the inequality:
$-2 \leq x \leq 3$

53. (Section 10.4)
$c \geq 2$

54. (Section 10.4)
a.

$$\begin{array}{cc} & \begin{array}{cc} H & T \end{array} \\ \begin{array}{c} H \\ T \end{array} & \begin{bmatrix} -4 & 6 \\ 6 & -3 \end{bmatrix} \end{array}$$

b. No, because no number is the smallest in its row and the largest in its column.

55. (Section 10.3)

a.

$$\begin{array}{cc} \text{Qz} & \text{No Qz} \end{array}$$

$$\begin{bmatrix} .2 & .8 \\ .5 & .5 \end{bmatrix} \begin{array}{l} \text{Qz} \\ \text{No Qz} \end{array}$$

b. Make "Qz" an absorbing state

$$\begin{array}{c|c} 1 & 0 \\ \hline .5 & .5 \end{array}$$

Then $T = (I - Q)^{-1} = (1 - .5)^{-1} = 2$
So 2 days.

56. (Section 10.3)

a.

$$\left[\begin{array}{cc|cc} 1 & 0 & 0 & 0 \\ 0 & 1 & 0 & 0 \\ \hline 0 & .1 & .9 & 0 \\ .7 & .3 & 0 & 0 \end{array}\right]$$

So $S = \begin{bmatrix} 0 & .1 \\ .7 & .3 \end{bmatrix}$

$Q = \begin{bmatrix} .9 & 0 \\ 0 & 0 \end{bmatrix}$

b. $\begin{bmatrix} 1 & 0 \\ 0 & 1 \end{bmatrix} - \begin{bmatrix} .9 & 0 \\ 0 & 0 \end{bmatrix} = \begin{bmatrix} .1 & 0 \\ 0 & 1 \end{bmatrix} = \begin{bmatrix} 10 & 0 \\ 0 & 1 \end{bmatrix}$

c. $\begin{bmatrix} 10 & 0 \\ 0 & 1 \end{bmatrix}\begin{bmatrix} 0 & .1 \\ .7 & .3 \end{bmatrix} = \begin{bmatrix} 0 & 1 \\ .7 & .3 \end{bmatrix}$

d. .7

57. (Section 10.3)
$T = [2]; T - S = [2][.3 \;\; .2] = [.6 \;\; .4]$

58. (Section 10.3)
$T = [1.25]; T - S = [1.25][.1 \;\; .7] = [.125 \;\; .875]$

59. (Section 10.3)

$T = \begin{bmatrix} 1.58 & .526 \\ 877 & 1.40 \end{bmatrix}$

$T \bullet S = \begin{bmatrix} 1.58 & .526 \\ .877 & 1.40 \end{bmatrix} \begin{bmatrix} .1 & .3 \\ .2 & .2 \end{bmatrix}$

$= \begin{bmatrix} .263 & .579 \\ .368 & .543 \end{bmatrix}$

60. (Section 10.3)

$T = \begin{bmatrix} 1.25 & .33 \\ .25 & 1.67 \end{bmatrix}$

$T \bullet S = \begin{bmatrix} 1.25 & .33 \\ .25 & 1.67 \end{bmatrix} \begin{bmatrix} 1/3 & 1/3 \\ 1/4 & 1/4 \end{bmatrix}$

$= \begin{bmatrix} .5 & .5 \\ .5 & .5 \end{bmatrix}$

61. (Section 10.3)
Absorbing

62. (Section 10.3)
Non-Absorbing

63. (Section 10.3)
Absorbing

64. Section 10.3)
Non-Absorbing

65. (Section 10.3)
Absorbing

66. (Section 10.3)
Non-Absorbing

67. (Section 10.3)
Absorbing

68. (Section 10.3)
Non-Absorbing

69. (Section 10.3)
Absorbing

70. (Section 10.3)
a. 1.5
b. 0.8
c. 0.3 + 1.8 + 2.1 = 4.2 games

71. (Section 10.2)

$[.6 \quad .4]\begin{bmatrix} .8 & .2 \\ .3 & .7 \end{bmatrix} = [.6 \quad .4]$; 60%

72. (Section 10.2)

$[.6 \quad .4]\begin{bmatrix} .8 & .2 \\ .3 & .7 \end{bmatrix} = [.6 \quad .4]$; 60%

73. (Section 10.2)
60%

74. (Section 10.2)

$[.7 \quad .3]\begin{bmatrix} .9 & .1 \\ .2 & .8 \end{bmatrix} = [.69 \quad .31]$

75. (Section 10.2)

$[.69 \quad .31]\begin{bmatrix} .9 & .1 \\ .2 & .8 \end{bmatrix} = [.683 \quad .317]$

76. (Section 10.2)

long range: $[\,x \quad 1 - x\,]\begin{bmatrix} .9 & .1 \\ .2 & .8 \end{bmatrix} = [\,x \quad 1 - x\,]$

solving: [2/3 1/3]

77. (Section 10.2)
.5

78. (Section 10.2)

$$\begin{array}{c} \\ B \\ Br \\ R \end{array}\begin{array}{c} \begin{array}{ccc} B & Br & R \end{array} \\ \begin{bmatrix} .5 & .3 & .2 \\ .3 & .6 & .1 \\ .2 & .6 & .2 \end{bmatrix} \end{array}\begin{bmatrix} .5 & .3 & .2 \\ .3 & .6 & .1 \\ .2 & .6 & .2 \end{bmatrix} = \begin{array}{c} B \\ Br \\ \end{array}\begin{array}{c} \begin{array}{ccc} B & & R \end{array} \\ \begin{bmatrix} .38 & .45 & .17 \\ .35 & .51 & .14 \\ .32 & .54 & .14 \end{bmatrix} \end{array}$$; 38%

79. (Section 10.2)

$$\begin{array}{c} \\ B \\ Br \\ R \end{array} \begin{array}{c} \begin{array}{ccc} B & Br & R \end{array} \\ \begin{bmatrix} .38 & .45 & .17 \\ .35 & .51 & .14 \\ .32 & .54 & .14 \end{bmatrix} \end{array} \begin{bmatrix} .5 & .3 & .2 \\ .3 & .6 & .1 \\ .2 & .6 & .2 \end{bmatrix} = B \begin{array}{c} B \\ \begin{bmatrix} .359 & .486 & .155 \\ .356 & .495 & .149 \\ .350 & .504 & .146 \end{bmatrix} \end{array} ; 36\%$$

80. (Section 10.2)
.6

81. (Section 10.2)
.14

82. (Section 10.2)
.356

83. (Section 10.2)

$$\begin{array}{ccc} B & Br & R \\ [.5 & .4 & .1] \end{array} \begin{array}{c} \begin{array}{ccc} B & Br & R \end{array} \\ \begin{bmatrix} .5 & .3 & .2 \\ .3 & .6 & .1 \\ .2 & .6 & .2 \end{bmatrix} \end{array} = \begin{array}{ccc} B & Br & R \\ [.39 & .45 & .16] \end{array}$$

84. (Section 10.2)

$$[.39 \quad .45 \quad .16] \begin{bmatrix} .5 & .3 & .2 \\ .3 & .6 & .1 \\ .2 & .6 & .2 \end{bmatrix} = \begin{array}{ccc} B & Br & R \\ [.362 & .483 & .155] \end{array}$$

85. (Section 10.2)

$$[.362 \quad .483 \quad .155] \begin{bmatrix} .5 & .3 & .2 \\ .3 & .6 & .1 \\ .2 & .6 & .2 \end{bmatrix} = \begin{array}{ccc} B & Br & R \\ [.3569 & .4914 & .1517] \end{array}$$

86. (Section 10.2)

Solving: $[x \quad y \quad 1-x-y] \begin{bmatrix} .5 & .3 & .2 \\ .3 & .6 & .1 \\ .2 & .6 & .2 \end{bmatrix} = [x \quad y \quad 1-x-y]$

Answer: [.356 .493 .151]

87. (Section 10.2)
8

88. (Section 10.3)
4

89. (Section 10.3)
4.5

90. (Section 10.3)
1/5

91. (Section 10.1)

$$\begin{bmatrix} .2 & .8 \\ .6 & .4 \end{bmatrix}$$

92. (Section 10.1)

$$[.5 \quad .5]\begin{bmatrix} .2 & .8 \\ .6 & .4 \end{bmatrix} = [.4 \quad .6]$$

.4

93. (Section 10.1)

$$[.5 \quad .5]\begin{bmatrix} .2 & .8 \\ .6 & .4 \end{bmatrix}\begin{bmatrix} .2 & .8 \\ .6 & .4 \end{bmatrix} = [.4 \quad .6]$$

.44

94. (Section 10.2)

$$[t \quad 1-t]\begin{bmatrix} .2 & .8 \\ .6 & .4 \end{bmatrix} = [t \quad 1-t]$$

$t = .429$

CHAPTER 11

LOGIC AND LOGIC CIRCUITS

In problems 1 – 4, determine which sentences are propositions.

1. Today is Friday.

2. What a great catch!

3. Who was Einstein?

4. The largest city in the U.S. is Sitka, Alaska.

In problems 5 – 8, negate each proposition.

5. Everyone loves finite mathematics.

6. Our text contains over 4000 exercises.

7. The stock market rose yesterday.

8. Some students do not study.

In problems 9 – 12, construct a truth table for each proposition.

9. $\sim p \vee q$

10. $\sim(\sim p \wedge \sim q)$

11. $(p \underline{\vee} q) \wedge p$

12. $(p \vee q) \wedge (\sim r)$

13. State the Absorption Laws.

14. State De Morgan's Laws.

In problems 15 – 17, write the converse, contrapositive, and inverse of each statement.

15. $p \Rightarrow q$

16. If today is Sunday, tomorrow is Monday.

17. It snows only if it is cold.

18. Construct a truth table for $\sim p \vee (p \Rightarrow q)$

19. Construct a truth table for $(p \vee q) \Rightarrow (p \wedge q)$

20. Construct a truth table for $(p \Rightarrow q) \Rightarrow (q \Rightarrow p)$

21. Prove the statement below using a direct proof.

    Only if stocks rise will gold fall.
    Bonds are rising.
    Either bonds are not rising or gold is falling.

    Prove: Stocks are rising.

22. Prove the statement below using an indirect proof.

    Only if stocks rise will bonds rise.
    Gold is falling.
    Bonds rise and gold falls.

    Prove: Stocks are rising.

23. Determine when the output of the circuit below is 1.

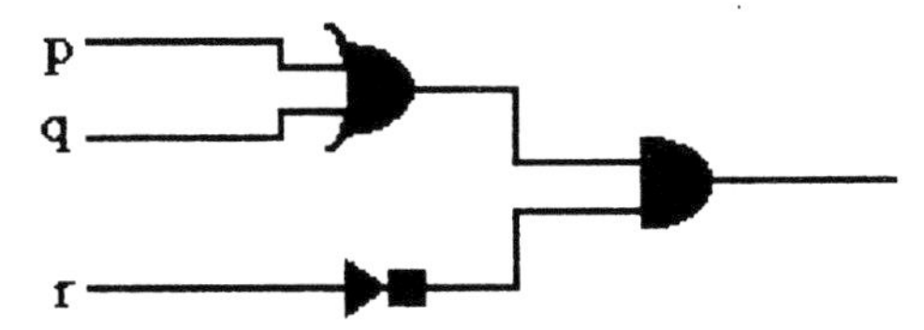

24. Determine when the output of the circuit below is 1.

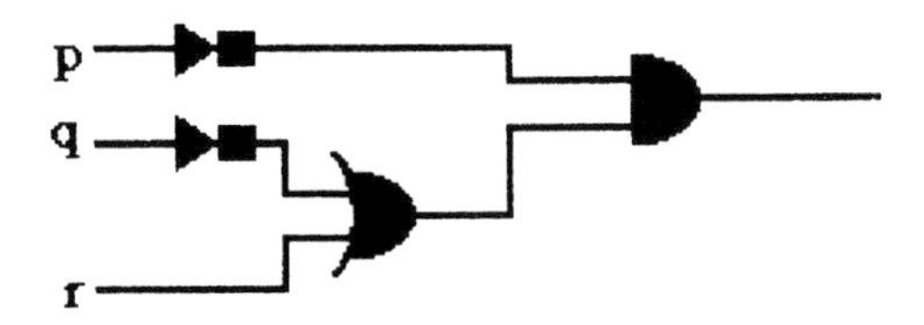

25. Show that the output of the circuit below is never 1.

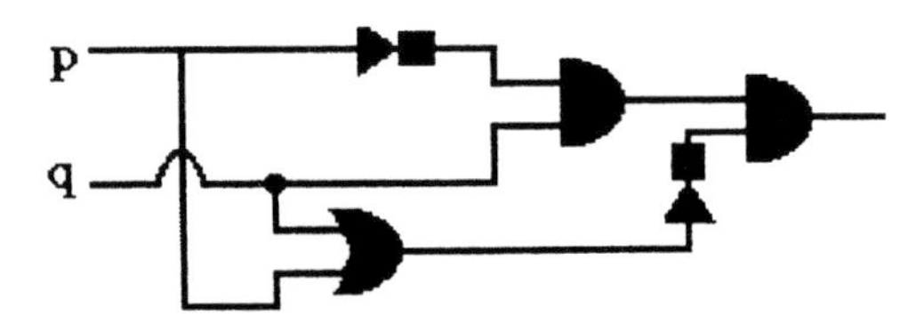

26. Simplify the circuit below.

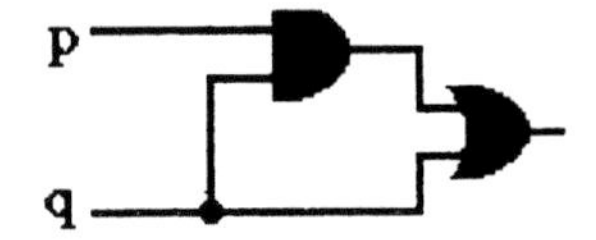

27. Construct a circuit corresponding to $\sim q \bullet \sim(p \oplus q)$

28. Which of the following is not logically equivalent to the others?

  a. All students write legibly.
  b. Each student writes legibly.
  c. Every medical student writes legibly.
  d. Any student writes legibly.

In problems 29 – 34, let p denote "Sue is smart" and let q denote "Sue likes math". Write a simple sentence for each of the following:

29. $p \vee q$

30. $\sim p$

31. $\sim p \wedge \sim q$

32. $p \underline{\vee} q$

33. $p \Rightarrow q$

34. $p \Leftrightarrow q$

35. Use De Morgan's Laws to negate the proposition "John is a great tennis player and is not talkative."

36. What does the symbol $\Leftrightarrow$ mean?

37. State the Law of Detachment.

38. What is another name for an indirect proof?

39. What is the name for the simplest logic circuits?

40. What are the three most basic types of gates?

41. Is $(p \Rightarrow q) \vee (q \Rightarrow p)$ a tautology?

42. Is $(p \vee q) \vee (\sim q)$ a tautology?

SOLUTIONS

1. (Section 11.1)
   Proposition

2. (Section 11.1)
   Not a Proposition

3. Section 11.1)
   Not a Proposition

4. (Section 11.1)
Proposition

5. (Section 11.1)
Someone does not love finite mathematics.

6. (Section 11.1)
Our text contains 4000 or fewer exercises.

7. (Section 11.1)
The stock market did not rise yesterday.

8. (Section 11.2)
Some students study.

9. (Section 11.2)

p	q	~p	~p ∨ q
T	T	F	T
T	F	F	F
F	T	T	T
F	F	T	T

10. (Section 11.2)

p	q	~p	~q	~p ∧ ~q	~(~p ∧ q)
T	T	F	F	F	T
T	F	F	T	F	T
F	T	T	F	F	T
F	F	T	T	T	F

11. (Section 11.2)

p	q	p $\underline{\vee}$ q	(p $\underline{\vee}$ q) ∧ p
T	T	F	F
T	F	T	T
F	T	T	F
F	F	F	F

12. (Section 11.2)

p	q	r	~r	p V q	(p V q) Λ (~r)
T	T	T	F	T	F
T	T	F	T	T	T
T	F	T	F	T	F
T	F	F	T	T	T
F	T	T	F	T	F
F	T	F	T	T	T
F	F	T	F	F	F
F	F	F	T	F	F

13. (Section 11.2)
$p \vee (p \wedge q) \equiv p$, $p \wedge (p \vee q) \equiv p$

14. (Section 11.2)
$\sim(p \vee q) \equiv \sim p \wedge \sim q$, $\sim(p \wedge q) \equiv \sim p \vee \sim q$

15. (Section 11.3)
converse: $q \Rightarrow p$
contrapositive: $\sim q \Rightarrow \sim p$
inverse: $\sim p \Rightarrow \sim q$

16. (Section 11.3)
converse: If tomorrow is Monday, then today is Sunday.
contrapositive: If tomorrow is not Monday, then today is not Sunday.
inverse: If today is not Sunday, then tomorrow is not Monday.

17. (Section 11.3)
converse: If it is cold, then it snows.
contrapositive: If it is not cold, then it will not snow.
inverse: If it does not snow, then it is not cold.

18. (Section 11.3)

p	q	~p	p ⇒ q	~p V (p ⇒ q)
T	T	F	T	T
T	F	F	F	F
F	T	T	T	T
F	F	T	T	T

19. (Section 11.3)

p	q	p ∨ q	p ∧ q	(p ∨ q) ⇒ (p ∧ q)
T	T	T	T	T
T	F	T	F	F
F	T	T	F	F
F	F	F	F	T

20. (Section 11.3)

p	q	p ⇒ q	q ⇒ p	(p ⇒ q) ⇒ (q ⇒ p)
T	T	T	T	T
T	F	F	T	T
F	T	T	F	F
F	F	T	T	T

21. (Section 11.4)
p: Stocks rise; q: Gold falls; r: Bonds rise

The following statements are true: q ⇒ p, r, (~r) ∨ q
Since r is true, ~r is false. Since (~r) ∨ q is true ~r is false, then q is true. Since q ⇒ p is true and q is true, then p is true.

22. (Section 11.4)
p: Stocks rise; q: Gold falls r: Bonds rise

The following statements are true: r ⇒ p, q, (~r) ∧ q
Assume the contrary, namely, that ~p is true. Since q is true and r ∧ q is true, then r is true. Since r is true and r ⇒ p is true, then p is true; but ~p is also true. This is Impossible. Thus p is true.

23. (Section 11.5)
We use a truth table:

p	q	r	p ⊕ q	~r	(p ⊕ q) r
1	1	1	1	0	0
1	1	0	1	1	1
1	0	1	1	0	0
1	0	0	1	1	1
0	1	1	1	0	0
0	1	0	1	1	1
0	0	1	0	0	0
0	0	0	0	1	0

24. (Section 11.5)
We use a truth table:

p	q	r	~p	~q	~q ⊕ r	~p ⊕ (~q ⊕ r)
1	1	1	0	0	1	0
1	1	0	0	0	0	0
1	0	1	0	1	1	0
1	0	0	0	1	1	0
0	1	1	1	0	1	1
0	1	0	1	0	0	0
0	0	1	1	1	1	1
0	0	0	1	1	1	1

25. (Section 11.5)
The output is

$$
\begin{aligned}
(\sim p \bullet q) \bullet [\sim(p \oplus q)] &= \sim p \bullet q \bullet \sim p \bullet \sim q \\
&= (\sim p \bullet \sim p) \bullet q \bullet (\sim q) \\
&= \sim p \bullet 0 \\
&= 0
\end{aligned}
$$

The output is never 1.

26. (Section 11.5)
The output is

$(p \bullet q) \oplus q = q$

A simpler design is

q ⟶

27. (Section 11.5)

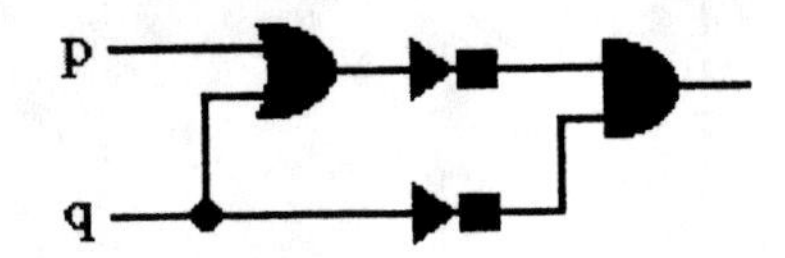

28. (Section 11.1)
c: Every medical student writes legibly.

29. (Section 11.1)
Sue is smart, or Sue likes math.

30. (Section 11.1)
Sue is not smart.

31. (Section 11.1)
Sue is not smart, and Sue does not like math.

32. (Section 11.1)
Sue is smart, or Sue likes math but not both.

33. (Section 11.3)
If Sue is smart, then she likes math.

34. (Section 11.3)
Sue is smart if and only if she likes math.

35. (Section 11.2)
Let p: John is a great tennis player.
q: John is talkative.

Our original statement becomes $p \wedge \sim q$.
Negating this becomes $\sim(p \wedge \sim q) \Rightarrow \sim p \vee q$ (by De Morgan's Laws).
So our solution is: John is not a great tennis player, or he is talkative.

36. (Section 11.3)
if and only if

37. (Section 11.4)
If both $p \Rightarrow q$ and p are true, then q must be true.

38. (Section 11.4)
proof by contradiction

39. (Section 11.5)
gate

40. (Section 11.5)
1. inverter (or not gate); 2. and; 3. or

41. (Section 11.3)

p	q	$p \Rightarrow q$	$q \Rightarrow p$	$(p \Rightarrow q) \vee (p \Rightarrow q)$
T	T	T	T	T
T	F	F	T	T
F	T	T	F	T
F	F	T	T	T

Yes.

42. (Section 11.3)

p	q	$p \vee q$	$\sim q$	$(p \vee q) \vee (\sim q)$
T	T	T	F	T
T	F	T	T	T
F	T	T	F	T
F	F	F	T	T

Yes.

CHAPTER 12

RELATIONS, FUNCTIONS, AND INDUCTION

1. Let A = {d, f, g, m, s, u}, B = {mother, father, son, daughter}.
   Define R to be the relation from A to B as: ${}_xR_y$ means y contains x.
   Write R as a set of ordered pairs.

2. Let A = {0, 1, 2}. Define R to be the relation on A as:
   ${}_aR_b$ means $(a + b) \in A$. Write R as a set of ordered pairs.

3. Show that the relation R on the set of real numbers defines as:
   ${}_aR_b$ means $a \geq b$ is reflexive and transitive.

4. Give an example of a relation on the set of real numbers which is transitive but neither reflexive nor symmetric.

5. Let f assign elements of A = {0, 1, 2} into elements of B = {a, b, c, d} as follows:
   f(0) = d, f(1) = b, f(2) = d. Does f define a function? Why or why not?

6. Given the sets A = {x, y, z} and B = {5}. Define a function f from A into B.

7. Let f be a function from the set of integers into the integers defined by f(n) = 3n.
   Is f one-to-one? Why or why not?

8. Consider the two sets A = {a, b, c} and B = {1, 2, 3, 4}. Can an onto function exist from
   (i) A into B? Explain.
   (ii) B into A? Explain.

9. Consider the sequence (s_n) defined by $s_n = \dfrac{n}{2n - 1}$.

   Write out s_0, s_1, s_2, s_{100}.

10. For the sequence (a_n) defined by $a_n = 2^{n-1}$, find a_0, a_1, a_2, a_3, a_4, a_{11}.

11. Let the sequence (b_n) be defined by $b_n = \dfrac{n}{2n + 3}$. Find $\dfrac{3}{2} b_5 - \dfrac{4}{3} b_4$.

12. Write a formula for the nth term of the sequence whose first few terms are 1, 1, 2, 6, 24, 120, 720, ...

13. Let S(n) be the statement $n < n + 1$

    (a) Write and tell if true: S(1), S(2), S(3), S(4)

    (b) Write S(k) and S(k +1).

14. Let S(n) be the statement

$$1 + 2 + 4 + 8 + 16 + \ldots + 2^n = 2^{n+1} - 1$$

(a) Write and tell if true: S(0), S(1), S(2), S(3), S(7)

(b) Write S(k) and S(k + 1)

15. Prove by mathematical induction the statement:

$$1 + 2 + 3 + 4 + \ldots + n = \frac{n(n+1)}{2}$$

16. Prove by mathematical induction the statement:

$$\frac{1}{1 \cdot 2} + \frac{1}{2 \cdot 3} + \frac{1}{3 \cdot 4} + \ldots + \frac{1}{n(n+1)} = \frac{n}{n+1}$$

17. Given the following recurrence relation and initial condition

$s_n = 2n + s_{n-1}$
$s_0 = 1$

Find s_1, s_2, s_3, s_4, and s_5

18. Given the following recurrence relation and initial conditions

$s_n = s_{n-1} - s_{n-2}$
$s_0 = 1$
$s_1 = 2$

Find s_2, s_3, s_4, s_5 and s_6

19. Consider the recurrence relation $s_n = 2s_{n-1} + 3$.

Find the initial condition s_0 if $s_1 = 1$.

20. State the reflexive property.

21. State the symmetric property.

22. State the transitive property.

23. Identify which property is used:

a. 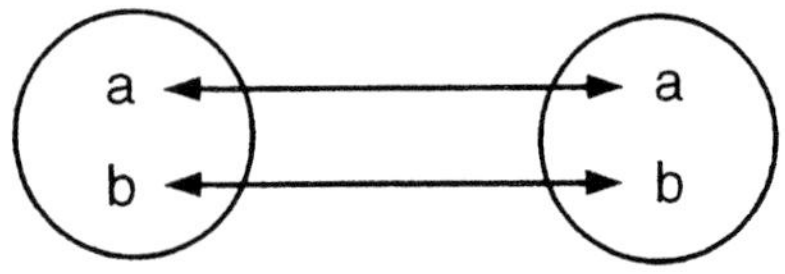 a ↔ a, b ↔ b

b. R = {(a, b), (b, a), (c, d), (d, c)}

c. If a = b and b = 3, then a = 3.

24. What is the name for a binary word of length 7?

25. $R = \{(2, 1), (3, 4), (6, 2), (8, 1)\}$
   a. What is R^{-1}?
   b. Is R a function?
   c. Is R^{-1} a function?
   d. Domain of R?
   e. Range of R?
   f. Domain of R^{-1}?

26. If $f(x) = 3x^2 - 2$, find
   a. f(4)
   b. f(0)
   c. f(−1)
   d. f(3a)

27. If H is the Hamming distance function, find H(1101, 1010)

28. Give the first four terms of the sequence defined by

$$S_N = (-1)^{N+1}\left(\frac{1}{N^3}\right),$$ where S_1 is the first term.

29. Give the next four terms of

$$S_N = (-1/2)^N S_{N-1}$$

if S_0 is 2.

30. $f(x) = \{(0, 1), (1, 2), (2, 3)\}$ and $g(x) = \{(1, 2), (2, 1), (3, 0)\}$. Find $f \circ g(1)$.

31. $f(x) = \{(0, 1), (1, 2), (2, 3)\}$ and $g(x) = \{(1, 2), (2, 1), (3, 0)\}$. Find $g \circ f(1)$.

SOLUTIONS

1. (Section 12.1)
R = {(d, daughter), (f, father), (g, daughter), (m, mother), (s, son), (u, daughter)}
Not unique.

2. (Section 12.1)
$R = \{(0, 0), (0, 1), (0, 2), (1, 0), (1, 1), (2, 0)\}$
Not unique.

3. (Section 12.1)
Let a, b, c be in R.
Since $a \geq a$ then ${}_aR_a$ thus R is reflexive.
And since $a \geq b$ and $b \geq c$ implies $a \geq c$ then R is transitive.

4. (Section 12.1)
The "greater than" relation.

5. (Section 12.2)
Yes, because every element in A is assigned a unique element in B.

6. (Section 12.2)
$f(x) = 5$, $f(y) = 5$, $f(z) = 5$.

7. (Section 12.2)
No, because the number 2 is not the image of any integer under this function.

8. (Section 12.2)
(a) No, because A has fewer elements than B.
(b) Yes. Consider f from B to A defined as: $f(1) = a$, $f(2) = b$, $f(3) = c$, $f(4) = c$.

9. (Section 12.3)
$$s_0 = \frac{0}{0-1} = 0;\ s_1 = \frac{1}{2-1} = 1;\ s_2 = \frac{2}{4-1} = \frac{2}{3};\ s_{100} = \frac{100}{200-1} = \frac{100}{199}.$$

10. (Section 12.3)

$a_0 = 2^{-1} = 1/2$; $a_1 = 2^0 = 1$; $a_2 = 2^1 = 2$; $a_3 = 2^2 = 4$; $a_4 = 2^3 = 8$; $a_{11} = 2^{10} = 1024$.

11. (Section 12.3)
$$b_5 = \frac{5}{10+3} = \frac{5}{13};\ b_4 = \frac{4}{8+3} = \frac{4}{11}.$$

Therefore, $$\frac{3}{2}b_5 - \frac{4}{3}b_4 = \frac{3}{2}\left(\frac{5}{13}\right) - \frac{4}{3}\left(\frac{4}{11}\right) = \frac{15}{26} - \frac{16}{33} = \frac{495-416}{858} = \frac{79}{858}$$

12. (Section 12.3)
$(n-1)!$

13. (Section 12.5)
(a) $S(1)$: $1 < 1 + 1$, true.
$S(2)$: $2 < 2 + 1$, true.
$S(3)$: $3 < 3 + 1$, true.
$S(4)$: $4 < 4 + 1$, true.

(b) $S(k)$: $k < k + 1$
$S(k + 1) = k + 1 < (k + 1) + 1$

14. (Section 12.5)
(a) $S(0)$: $= 1 = 2^1 - 1$, true.
$S(1)$: $= 1 + 2 = 2^2 - 1$, true.
$S(2)$: $= 1 + 2 + 4 = 2^3 - 1$, true.
$S(3)$: $= 1 + 2 + 4 + 8 = 2^4 - 1$, true.
$S(7)$: $= 1 + 2 + 4 + 8 + 16 + 64 + 128 = 2^8 - 1$, true.

(b) $S(k)$: $1 + 2 + 4 + 8 + 16 + \ldots + 2^k = 2^{k+1} - 1$
$S(k + 1)$: $1 + 2 + 4 + 8 + 16 + \ldots + 2^k + 2^{k+1} = 2^{(k+1)+1} - 1$.

15. (Section 12.5)

Let S(n) be the statement $1 + 2 + 3 + 4 + ... + n = \frac{n(n + 1)}{2}$.

S(1) is $1 = \frac{1(1 + 1)}{2}$ which is true.

Assume S(k): $1 + 2 + 4 + ... + k = \frac{k(k + 1)}{2}$ is true.

Show that S(k + 1) is also true. Now S(k + 1) is

$$1 + 2 + 3 + 4 + ... + k + (k + 1) = \frac{(k + 1)[(k + 1) + 1]}{2}$$

Using our assumption, the left hand side gives

$$\frac{k(k + 1)}{2} + (k + 1) = \frac{k(k + 1) + 2(k + 1)}{2} = \frac{(k + 1)(k + 2)}{2} = \frac{(k + 1)[(k + 1) + 1]}{2}$$

which is the right hand side of S(k + 1). Thus S(k + 1) is true.

16. (Section 12.5)

Let S(n) be the statement $\frac{1}{1 \cdot 2} + \frac{1}{2 \cdot 3} + \frac{1}{3 \cdot 4} + ... + \frac{1}{n(n + 1)} = \frac{n}{n + 1}$

S(1) is $\frac{1}{1 \cdot 2} = \frac{1}{1 + 1}$ which is true.

Assume S(k): $\frac{1}{1 \cdot 2} + \frac{1}{2 \cdot 3} + \frac{1}{3 \cdot 4} + ... + \frac{1}{k(k + 1)} = \frac{k}{k + 1}$ is true.

Show that S(k + 1) is also true. Now S(k + 1) is

$$\frac{1}{1 \cdot 2} + \frac{1}{2 \cdot 3} + ... + \frac{1}{k(k + 1)} + \frac{1}{(k + 1)(k + 1) + 1} = \frac{k + 1}{(k + 1) + 1}$$

Left hand side is (using our assumption):

$$\frac{k}{k + 1} + \frac{1}{(k + 1)[(k + 1) + 1]} = \frac{k(k + 2) + 1}{(k + 1)(k + 2)} + \frac{k^2 + 2k + 1}{(k + 1)(k + 2)} = \frac{(k + 1)(k + 1)}{(k + 1)(k + 2)} = \frac{k + 1}{k + 2}$$

which is the right hand side of S(k + 1). Thus S(k + 1) is true.

17. (Section 12.6)

$s_1 = 2(1) + s_0 = 2 + 1 = 3$
$s_2 = 2(2) + s_1 = 4 + 3 = 7$
$s_3 = 2(3) + s_2 = 6 + 7 = 13$
$s_4 = 2(4) + s_3 = 8 + 13 = 21$
$s_5 = 2(5) + s_4 = 10 + 21 = 31$

18. (Section 12.6)
$s_2 = s_1 - s_0 = 2 - 1 = 1$
$s_3 = s_2 - s_1 = 1 - 2 = -1$
$s_4 = s_3 - s_2 = -1 - 1 = -2$
$s_5 = s_4 - s_3 = -2 + 1 = -1$
$s_6 = s_5 - s_4 = -1 + 2 = 1$

19. (Section 12.6)
$s_1 = 2s_0 + 3$ that is $1 = 2s_0 + 3$. Thus $s_0 = -1$

20. (Section 12.1)
For every $a \in A$, ${}_aR_a$.

21. (Section 12.1)
For every $a, b \in A$, If ${}_aR_b$ then ${}_bR_a$.

22. (Section 12.1)
For every $a, b, c, \in A$. If ${}_aR_b$ and ${}_bR_c$ then ${}_aR_c$.

23. (Section 12.1)
a. reflexive
b. symmetric
c. transitive

24. (Section 12.1)
byte

25. (Section 12.1)
a. $R^{-1} = \{(1, 2), (4, 3), (2, 6), (1, 8)\}$

(Section 12.2)
b. yes
c. no
d. Domain of $R = \{2, 3, 6, 8\}$
e. Range of $R = \{1, 2, 4\}$
f. Domain of $R^{-1} = \{1, 2, 4\}$

26. (Section 12.2)
a. $f(4) = 3 \bullet 4^2 - 2 = 3(16) - 2 = 48 - 2 = 46$
b. $f(0) = 3 \bullet 0^2 - 2 = 0 - 2 = -2$
c. $f(-1) = 3(-1)^2 - 2 = 3 - 2 = 1$
d. $f(3a) = 3 \bullet (3a)^2 - 2 = 3 \bullet 9a^2 - 2 = 27a^2 - 2$

27. (Section 12.2)
3

28. (Section 12.3)

$$S_1 = (-1)^{1+1}\left(\frac{1}{1^3}\right) = 1$$

$$S_2 = (-1)^3\left(\frac{1}{2^3}\right) = \frac{-1}{8}$$

$$S_2 = (-1)^4\left(\frac{1}{3^3}\right) = \frac{1}{27}$$

$$S_2 = (-1)^5\left(\frac{1}{4^3}\right) = \frac{-1}{256}$$

29. (Section 12.5)

If $S_0 = 2$, then $S_1 = \left(\frac{-1}{2}\right)^1 \cdot 2 = -1$

$$S_2 = \left(\frac{-1}{2}\right)^2 \cdot (-1) = \frac{-1}{4}$$

$$S_2 = \left(\frac{-1}{2}\right)^3 \cdot \left(\frac{-1}{4}\right) = \frac{1}{32}$$

$$S_2 = \left(\frac{-1}{2}\right)^4 \cdot \left(\frac{1}{32}\right) = \frac{1}{512}$$

30. (Section 12.2)
1

31. (Section 12.2)
3

CHAPTER 13

GRAPHS AND TREES

1. Draw a simple graph that has three vertices where each vertex is of degree 2.

2. Draw a simple graph that has five vertices where each vertex is of degree 4.

3. Does a graph with vertices v_1, v_2, v_3, and v_4 of degrees 2, 1, 1, and 3, respectively, exist? Explain.

4. What is the minimum number of edges needed to join nine vertices to form a simple graph where each vertex is of degree at least 1.

5. For the graph shown at right, find, starting at v_1,
   (a) a path that is not simple and contains all edges.
   (b) a simple path that contains e_4.
   (c) a simple circuit.
   (d) the degree of each vertex.

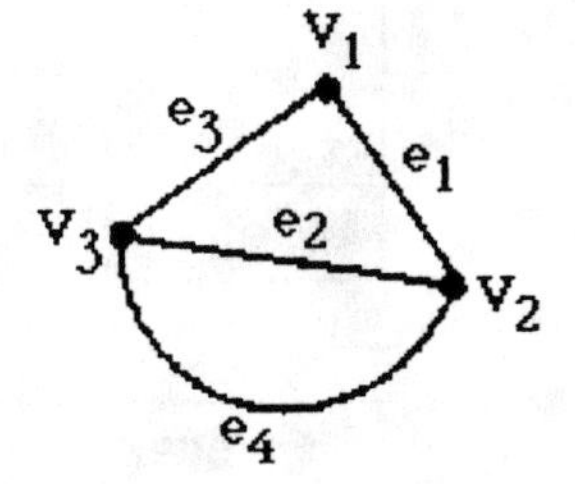

6. For the graph shown at right, does there exist
   (a) a simple path from v_1 to v_4 containing v_3? Explain.
   (b) a simple circuit starting at v_3 and containing v_1? Explain.

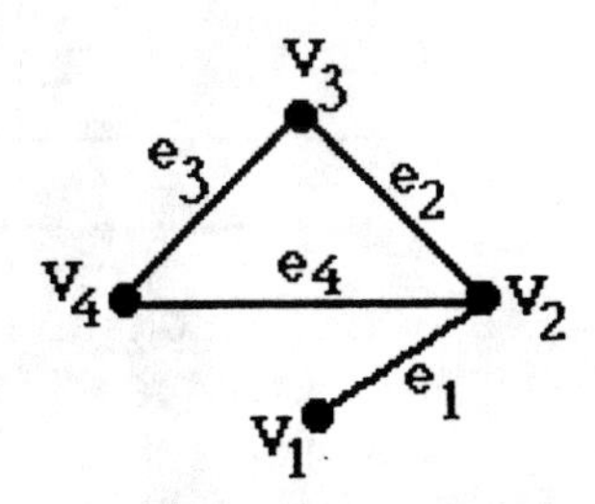

7. Draw a simple circuit consisting of three vertices.

8. Can a graph with 6 vertices and 4 edges be connected? Explain.

9. Which of the following two graphs contains an Eulerian circuit? Explain.

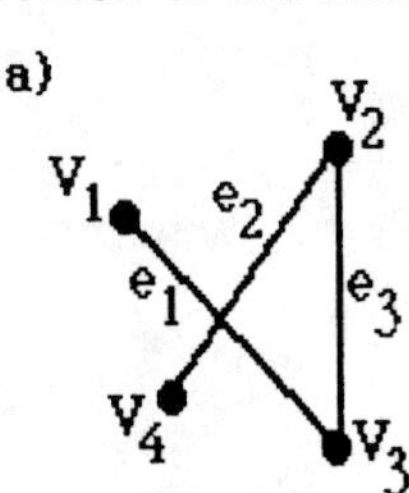

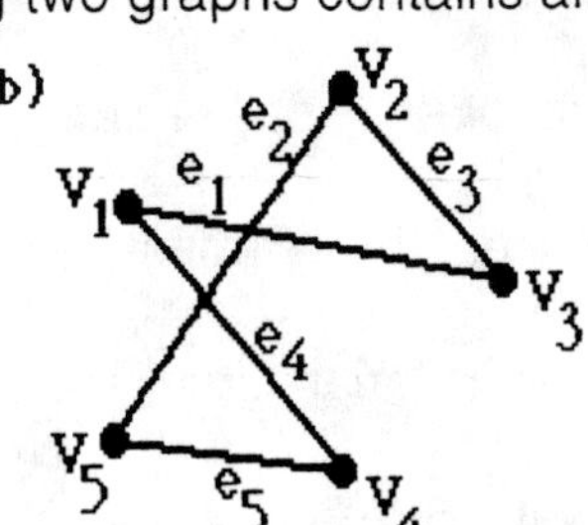

10. Draw a graph that (a) has four vertices and (b) contains an Eulerian circuit. Justify your answer.

11. Draw a graph that contains an Eulerian circuit but not a Hamiltonian circuit.

12. Give an example of a graph with 5 vertices that contains a Hamiltonian circuit but not an Eulerian circuit.

13. Draw a binary tree with seven vertices.

14. Draw a binary tree that has five vertices, two leaves, and two internal vertices.

15. Represent the following algebraic expression with a binary tree:

    $(a + b/c) - [(d + 5) - 2a]$

16. (a) Construct a binary search tree that stores the following set in a decreasing order 1, 9, 6, 2, 8, 3
    (b) Add the number 5 to the tree in (a)

17. Draw a digraph with the following properties:

Arc	Initial Point	Terminal Point
e_1	v_1	v_2
e_2	v_2	v_3
e_3	v_1	v_3
e_4	v_3	v_2

18. Draw the digraph with the following specifications:

Vertex	Indegree	Outdegree
v_1	0	1
v_2	1	1
v_3	2	1

19. Draw a digraph with 4 vertices v_1, v_2, v_3, and v_4 where
    v_1 is not reachable from v_2, v_3, or v_4
    v_2 is reachable from v_1, v_3, and v_4
    v_3 is reachable from v_1, v_2, and v_4
    v_4 is reachable from v_1, v_2, or v_3

20. Give an example of a connected graph that cannot be orientable.

21. A graph with no parallel edges and no loops is called a(n) _________.

22. If P is a connected graph with 12 vertices, then P must have at least how many edges?

23. What are the two criteria for S to be an Eulerian circuit.

24. If N is a simple connected graph with 6 vertices and N is a Hamiltonian circuit with non-adjacent vertices v_1 and v_3, then what must be true about $\deg(v_1) + \deg(v_3)$?

25. L is a tree. What two things must be true about L?

26. Represent the following algebraic expression with a binary tree. $(4 + x)(3 - y)$

27. Draw an Eulerian circuit with 4 vertices and 5 edges.

28. Is it possible to draw an Eulerian circuit with 4 vertices, 5 edges, and no loops?

SOLUTIONS

1. Section 13.1)

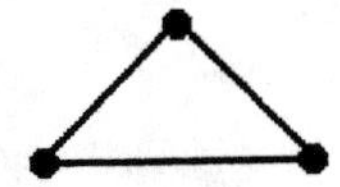

2. (Section 13.1)

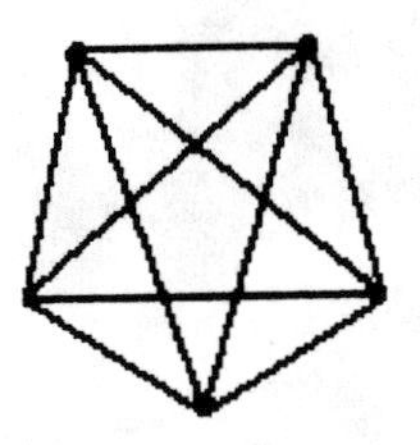

3. Section 13.1)
No, because the sum of degree of vertices is odd (=7).

4. (Section 13.1)
Five,

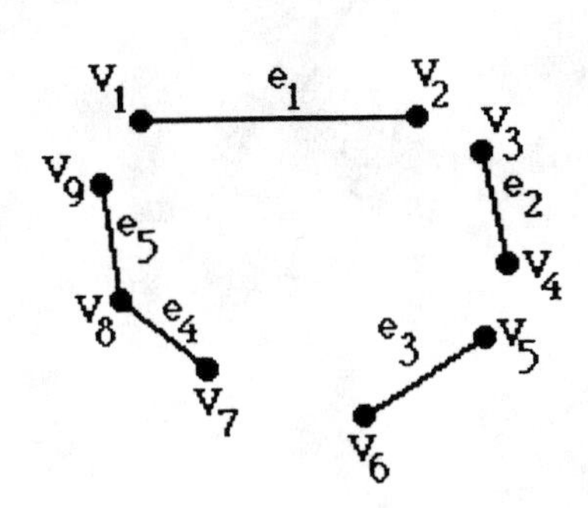

5. (Section 13.2)
(a) $v_2e_2v_3e_3v_1e_1v_2e_4v_3$
(b) $v_1e_1v_2e_4v_3$
(c) $v_1e_1v_2e_2v_3e_3v_1$
(d) $\deg(v_1) = 2$ $\deg(v_2) = 3$ $\deg(v_3) = 3$

6. (Section 13.2)
(a) Yes. Consider the simple path $v_1e_1v_2e_2v_3e_3v_4$.
(b) No. Any circuit containing v_1 must contain e_1 twice, thus no such simple circuit exists.

7. (Section 13.2)
The simple circuit is $v_1e_1v_2e_2v_3e_3v_1$.

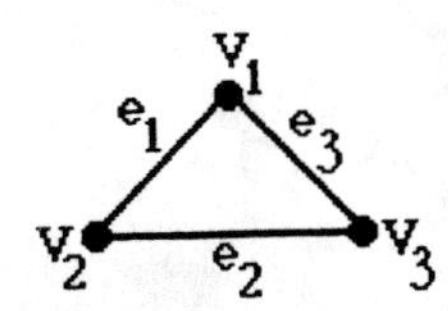

8. (Section 13.2)
No. At least 5 edges are needed.

9. (Section 13.3)
Graph (a) has vertex v_1 of odd degree; therefore, it contains no Eulerian circuit
Graph (b) contains the following Eulerian circuit: $v_1e_1v_3e_3v_2e_2v_5e_5v_4e_4v_1$.

10. (Section 13.3)

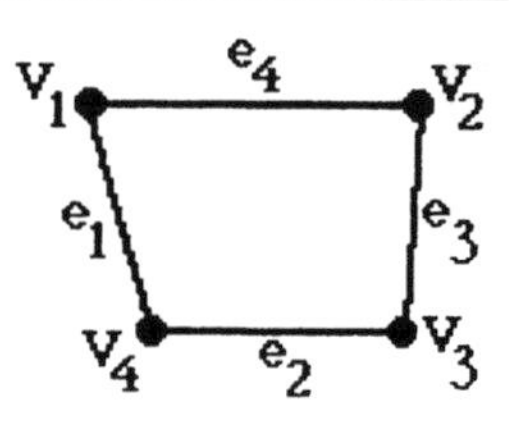

Eulerian circuit: $v_1e_1v_2e_2v_3e_3v_4e_4v_1$

11. (Section 13.3)

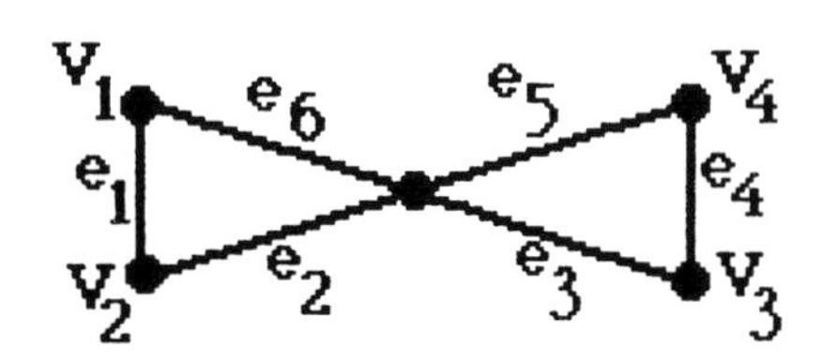

12. (Section 13.3)

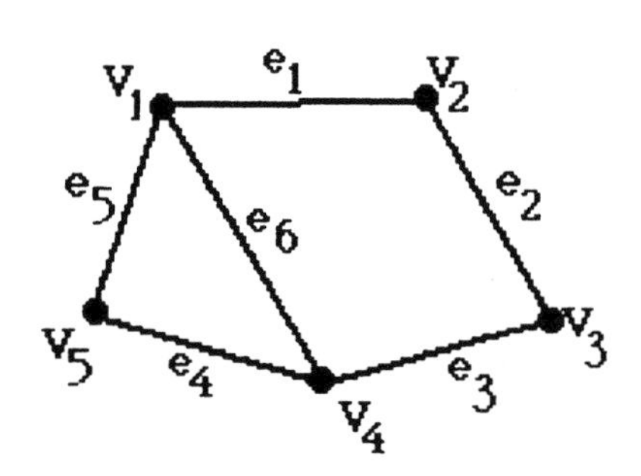

13. (Section 13.4)

14. (Section 13.4)

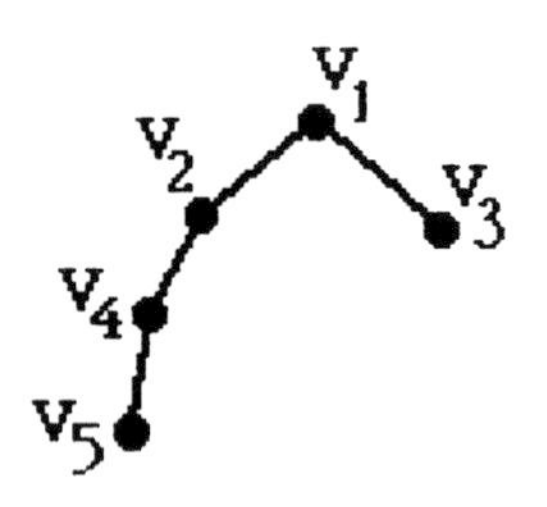

Internal vertices: v_2 and v_4.
Leaves: v_3 and v_5.

15. (Section 13.4)

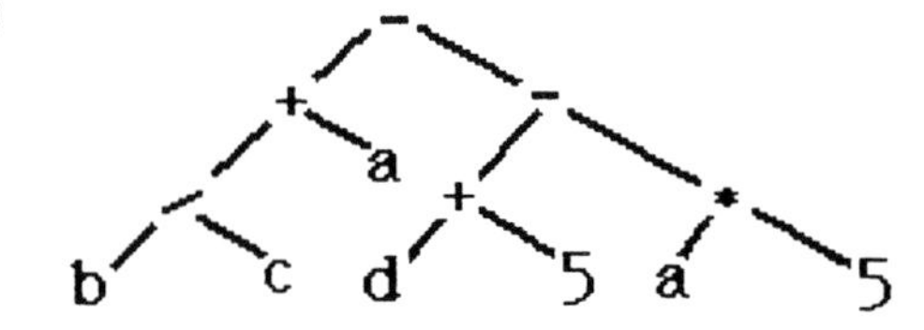

16. (Section 13.4)

a.

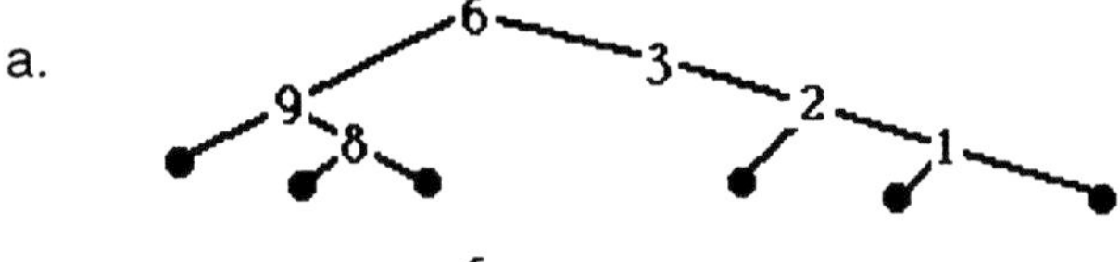

b.

17. (Section 13.5)

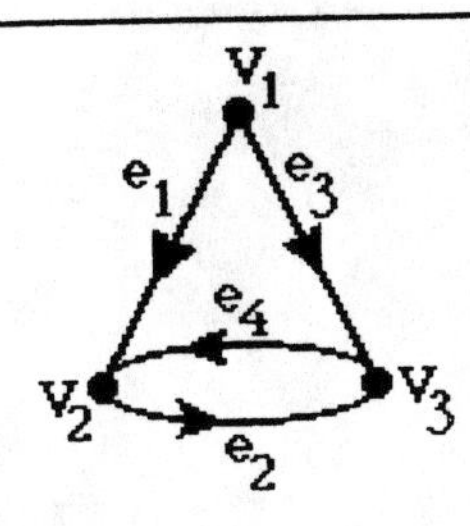

18. (Section 13.5)

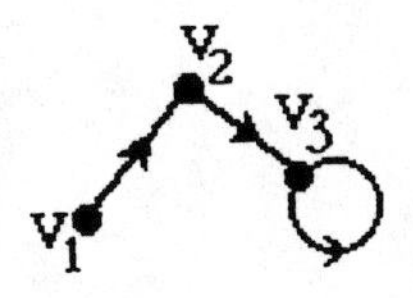

19. (Section 13.5)

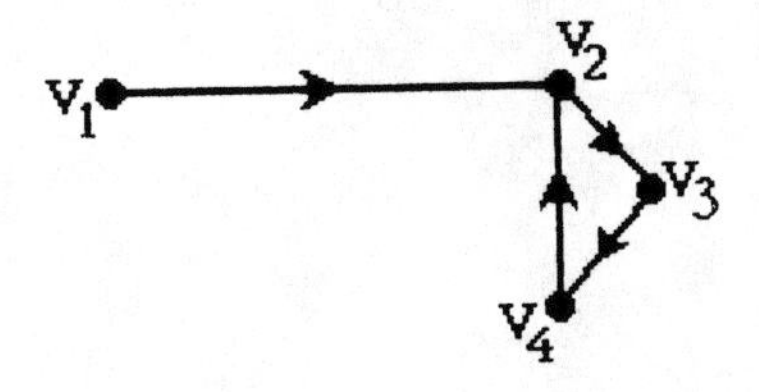

20. (Section 13.5)
Such a graph must contain a bridge:

21. (Section 13.1)
simple graph

22. (Section 13.2)
11

23. (Section 13.3)
1. S is connected.
2. Every vertex if S has an even degree.

24. (Section 13.3)
$\deg(v_1) + \deg(v_2) \geq 6$.

25. (Section 13.4)
1. L is connected.
2. L contains no non-trivial circuits.

26. (Section 13.4)

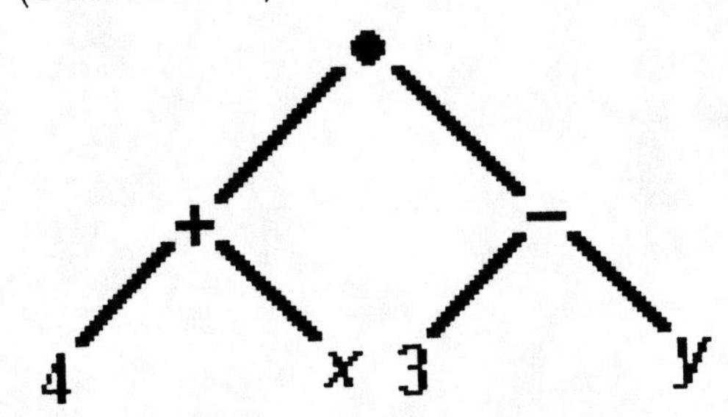

27. (Section 13.3)

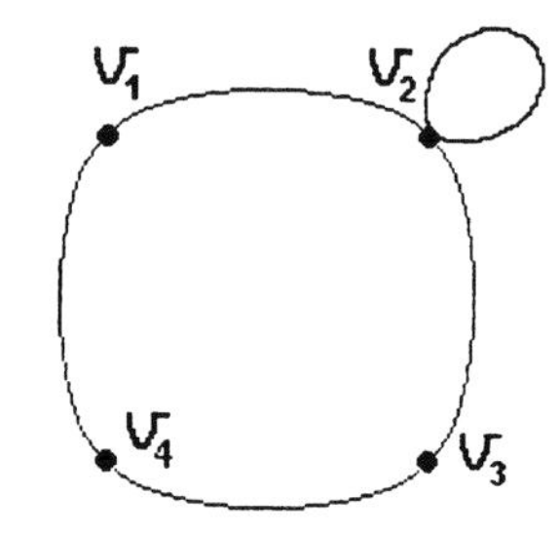

28. (Section 13.3)
No.